Springer Proceedings in Physics

Volume 260

Indexed by Scopus

The series Springer Proceedings in Physics, founded in 1984, is devoted to timely reports of state-of-the-art developments in physics and related sciences. Typically based on material presented at conferences, workshops and similar scientific meetings, volumes published in this series will constitute a comprehensive up-to-date source of reference on a field or subfield of relevance in contemporary physics. Proposals must include the following:

- name, place and date of the scientific meeting
- a link to the committees (local organization, international advisors etc.)
- scientific description of the meeting
- list of invited/plenary speakers
- an estimate of the planned proceedings book parameters (number of pages/ articles, requested number of bulk copies, submission deadline).

Please contact:

For Americas and Europe: Dr. Zachary Evenson; zachary.evenson@springer.com

For Asia, Australia and New Zealand: Dr. Loyola DSilva; loyola.dsilva@springer.com

More information about this series at http://www.springer.com/series/361

Stelio Montebugnoli · Andrea Melis ·
Nicolò Antonietti
Editors

The Search for ExtraTerrestrial Intelligence

Proceedings of the 2nd SETI-INAF Meeting 2019

Springer

Editors
Stelio Montebugnoli
IRA
INAF
Bologna, Italy

Andrea Melis
Osservatorio Astronomico di Cagliari
INAF
Selargius, Italy

Nicolò Antonietti
IRA-INAF
Bologna, Italy

ISSN 0930-8989   ISSN 1867-4941 (electronic)
Springer Proceedings in Physics
ISBN 978-3-030-63805-4   ISBN 978-3-030-63806-1 (eBook)
https://doi.org/10.1007/978-3-030-63806-1

This Springer imprint is published by the registered company Springer Nature Switzerland AG
The registered company address is: Gewerbestrasse 11, 6330 Cham, Switzerland

If we were to discover the existence of ***other civilizations****, we would not be able to talk to them, but our culture would take a giant leap forward.*
Nichi D'Amico (1953 - 2020)

Foreword

The fascinating SETI—Search for Extra Terrestrial Intelligence—program is experiencing renewed worldwide interest due to several favorable circumstances. From a scientific point of view, the discovery of thousands of planets outside the Solar System in just a tiny portion of our Galaxy, the breathtaking results on the atmosphere and geology of Mars and the progress in astrobiology encourage us to revisit the SETI objectives. At the same time, the improved parameter space of the most advanced astronomical facilities in all bands of the spectrum provides an enormous increase in the potential of current observations, which is essential for the SETI project.

In September 2019, the Italian astronomical community met in Rome to discuss the state of the art of SETI research and its perspectives, in light of the current and forthcoming scientific and technological landscape.

Only one year after this Second INAF SETI Meeting, our community unexpectedly lost its President, Prof. Nichi D'Amico. With his inclusive and open-minded approach to science, which we will all miss, Nichi has always supported SETI research, and recognized its relevance among INAF activities. This volume is gratefully dedicated to his memory.

Bologna, Italy

Tiziana Venturi
Director of Institute of Radio Astronomy-INAF

SRT
SARDINIA
RADIO
TELESCOPE
Foto Paolo Soletta INAF-OAC 2020

Preface

The discoveries of many new Earth-like planets at worldwide radio astronomical and optical observatories, day after day, have made the SETI program increasingly popular. In addition to the technological aspect, this program involves different approaches from a scientific point of view and, at various levels, many branches of human knowledge, such as philosophy, psychology, and many others. Today, humankind seems to expect a signal from the heavens from an intelligent extraterrestrial civilization, perhaps looking for someone who will be able to help it face its weaknesses and frailties.

Two years after the previous October 2017 SETI-INAF meeting, a second edition was held at the INAF headquarters in Rome. Speakers presented the national capabilities, potentiality, and instruments in the search for artificial radio and optical signals. This program has a promising future based on new approaches, newfound knowledge, and not least, state-of-the-art technology. Our late President Prof. Nicolò D'Amico, a strong supporter of the SETI program at INAF, stated that a possible detection of an ETI signal could represent one of the most important discoveries for the culture of humankind. Due to the incommensurable distances, it would not be possible to communicate with them, but it would be fundamental to understand that we are not alone!

Bologna, Italy — Stelio Montebugnoli
Selargius, Italy — Andrea Melis
Bologna, Italy — Nicolò Antonietti

Acknowledgements

We would like to thank the many people who have made this SETI meeting possible and successful.

Firstly, we thank the INAF President Nichi D'Amico (RIP): without him, the SETI meeting and this volume would not have been possible.

We also thank the Scientific Directorate for the considerable economic and logistical support.

Special thanks to the IRA Director Tiziana Venturi, for her official welcome and opening of the meeting, as well as for the writing of a lovely foreword.

We thank all of the participants and speakers and, in particular, the three chairs of the sessions: Daniela De Paulis, Francesco Schillirò, and Silvia Casu. We also thank Stefania Varano for coordinating the final discussion about outreach activities, which is very important in the SETI field.

Special thanks to Federico Gualano, who managed the entire organizational and administrative aspect.

Thanks to Raimondo Concu and Paolo Soletta for their essential technical and logistical support before and during the meeting; Paolo Soletta has also created the three images that are shown in the preface and on the front cover. We are also grateful to Chiara Giorgieri for providing a video of the meeting (available at the following link: https://www.youtube.com/watch?v=Cc_pQgiHF2Q&t=21950s) and to Simone Mattana for the audio and video support.

Special thanks to one of the most influential people in SETI history, Prof. Ivan Almar (Konkoly Observatory, Budapest), for attending the meeting and giving an excellent presentation.

Last but not least, a big thank you to Tiziana Coiana for handling all of the administrative practices needed to publish this volume, in collaboration with Shalini Monica Clement Selvam and Marina Forlizzi of Springer, to whom we extend our sincere gratitude.

Stelio Montebugnoli
Andrea Melis
Nicolò Antonietti

SOC, LOC and Participants

SOC

Stelio Montebugnoli
Andrea Melis
Raimondo Concu

LOC

Federico Gualano
Andrea Melis
Raimondo Concu
Paolo Soletta

Participants

Ivan Almar
Aldo Altamore
Nicolò Antonietti
Amedeo Balbi
Davide Bardelli
Germano Bianchi
Caterina Boccato
Alberto Buzzoni
Patrizia Caraveo
Pietro Cassaro
Silvia Casu
Cesare Cecchi Pestellini
Stefano Ciprini

Raimondo Concu
Daniela De Paulis
Massimo Della Valle
Marco Sergio Erculiani
Niccolò Focardi
Mauro Ghigo
Jarred Green
Roberto Lulli
Claudio Maccone
Michele Maris
Sabrina Masiero
Andrea Melis
Giuseppe Mogavero
Emilio Molinari
Stelio Montebugnoli
Marco Montuori
Silvia Motta
Roberto Orosei
Pierpaolo Pari
Delphine Perrodin
Vladimir Philippov
Salvatore Pluchino
Andrea Possenti
Arianna Ricchiuti
Giuseppe Savio
Francesco Schillirò
Roberto Silvotti
Paolo Soletta
Grazia Umana
Stefania Varano
Carlo Venturi
Tiziana Venturi
Giampaolo Vettolani
Giovanni Vladilo

Contents

Contributors

Nicolò Antonietti INAF Istituto di Radio Astronomia, Bologna, Italy

Amedeo Balbi Dipartimento di Fisica, Università di Roma "Tor Vergata", Roma, Italy

Caterina Boccato European Astrobiology Institute, La Palma, Spain;
Istituto Nazionale Di Astrofisica, INAF, Roma, Italy

Marta Burgay INAF Osservatorio Astronomico di Cagliari, Selargius, Italy

Patrizia A. Caraveo IASF-INAF Via A.Corti 12, Milan, Italy

Cesare Cecchi-Pestellini INAF - Osservatorio Astronomico di Palermo, Sicily, Italy

Milan M. Ćirković Astronomical Observatory of Belgrade, Belgrade, Serbia;
Faculty of Philosophy, Future of Humanity Institute, University of Oxford, Oxford, UK

Raimondo Concu INAF Osservatorio Astronomico di Cagliari, Selargius, Italy

Daniela de Paulis IAA SETI Permanent Committee, Roma, Italy

Marco Sergio Erculiani IRA-Istituto di Radioastronomia, Italian National Institute for Astrophysics (INAF), Bologna, Italy

Roberto Lulli National Institute for Astro-Physics (INAF), Radio Astronomy Institute (IRA), Bologna, Italy;
School of Advanced Studies (SAS), Camerino University, Camerino, Italy

Claudio Maccone INAF Osservatorio Astronomico di Cagliari, Selargius, Italy;
International Academy of Astronautics (IAA) and Istituto Nazionale di Astrofisica (INAF), Torino (Turin), Italy

Michele Maris INAF - Osservatorio Astronomico di Trieste, Trieste, Italy

Sabrina Masiero GAL Hassin Foundation - International Center for Astronomical Sciences, Isnello (PA), Italy;
Palermo Astronomical Observatory, Italian National Institute for Astrophysics (INAF), Palermo, Italy

Andrea Melis INAF Osservatorio Astronomico di Cagliari, Selargius, Italy

Emilio Molinari INAF Osservatorio Astronomico di Cagliari, Selargius, Italy

Stelio Montebugnoli INAF Istituto di Radio Astronomia, Bologna, Italy;
National Institute for Astro-Physics (INAF), Radio Astronomy Institute (IRA), former engineer in charge of the Northern Cross Radio Telescope, Medicina, Italy;
National Institute for Astro-Physics (INAF), Scientific Directorate, SETI advisor, Rome, Italy

Giuseppe Murante INAF - Osservatorio Astronomico di Trieste, Trieste, Italy

Alessandro Navarrini INAF Osservatorio Astronomico di Cagliari, Selargius, Italy

Julie Nekola Nováková European Astrobiology Institute, La Palma, Spain;
Charles University, Prague, Czech Republic

Roberto Orosei Istituto Nazionale di Astrofisica, Istituto di Radioastronomia, Bologna, Italy

Pierpaolo Pari INAF Istituto di Radio Astronomia, Bologna, Italy;
INAF - Istituto di Radio Astronomia, Bologna, Italy

Delphine Perrodin INAF Osservatorio Astronomico di Cagliari, Selargius, Italy

Maura Pilia INAF Osservatorio Astronomico di Cagliari, Selargius, Italy

Andrea Possenti INAF Osservatorio Astronomico di Cagliari, Selargius, Italy

Giuseppe Pupillo National Institute for Astro-Physics (INAF), Radio Astronomy Institute (IRA), Bologna, Italy

F. Schillirò INAF-Osservatorio Astrofisico di Catania, Catania, Italy

Laura Silva INAF - Osservatorio Astronomico di Trieste, Trieste, Italy

Paolo Simonetti Università degli Studi di Trieste, Trieste, Italy

G. Umana INAF-Osservatorio Astrofisico di Catania, Catania, Italy

Giovanni Vladilo INAF - Osservatorio Astronomico di Trieste, Trieste, Italy

Chapter 1
KLT, A New Algorithm For SETI

Nicolò Antonietti and Pierpaolo Pari

Abstract In 1960, the astronomer Frank Drake listened to the stars Tau Ceti and Epsilon Eridani, searching for extra terrestrial intelligent signals: the so called "Ozma project" was the first experimental SETI attempt. Since then, SETI has been mainly radio, although optical SETI and the technosignature search are becoming more and more of interest. Several methods exists in order to process radio signals and the KLT—Karhunen Loève Transform—is one of those. First in 1983 at the Nancy Radiotelescope, second at Ohio State University respectively Francois Biraud and Robert Dixon tried to use it recover very feeble signals buried into noise. In these proceedings, we analyze a new technique for KLT.

1.1 Properties of the KLT

The KLT is a mathematical tool that expands a stochastic process $X(t)$ into the convergent infinite series (1.1) (see [1])

$$X(t) = \sum_{n=1}^{\infty} Z_n \phi_n(t), \tag{1.1}$$

called the Karhunen-Loève transform, or KLT.
The time ranges in the closed interval $t = [0, T]$.
The Z_n's are random variables and are orthogonal, that is

$$E\{Z_m Z_n\} = \delta_{mn} \lambda_n \tag{1.2}$$

The $\phi_n(t)$ functions embed the whole dependence on time.

N. Antonietti · P. Pari (✉)
INAF - Istituto di Radio Astronomia, via Gobetti 101, 40129 Bologna, Italy
e-mail: pari.pierpaolo@gmail.com

S. Montebugnoli and A. Melis (eds.), *The Search for ExtraTerrestrial Intelligence*, Springer Proceedings in Physics 260, https://doi.org/10.1007/978-3-030-63806-1_1

The kernel of the KL transform is the autocorrelation; in fact, the $\phi_n(t)$ functions are the eigenfunctions of the autocorrelation, according to the integral equation (1.3)

$$\int_0^T E\{X(t_1)X(t_2)\}\,\phi_n(t_2)\mathrm{d}t_2 = \lambda_n\phi_n(t_1) \tag{1.3}$$

The unbiased estimator of the covariance of a stochastic process $X(t)$ have the matrix elements C_{ij} in (1.4) (see [2]):

$$C_{ij} = \frac{1}{M-1}\sum_{\alpha=0}^{M-1}(x_{\alpha i} - \mu_i)(x_{\alpha j}\mu_j)^* \tag{1.4}$$

being M the samples of the signal and μ_i the mean of the ith realization (random variable) of the stochastic process.
The autocorrelation of the stochastic process is given by (1.5)

$$R_{ij} = \frac{C_{ij}}{\sigma_i\sigma_j} \tag{1.5}$$

where σ_i is the standard deviation of the ith realization (random variable) of the stochastic process.

When the process is wide sense stationary, then the matrix is simpler as it becomes Toeplitz.

1.2 Algorithm

1.2.1 Features

Since the beginning, digital orthogonal transforms have been widely used to process data from speech, seismic, radar and astronomy fields. One of the most used tool to analyze a signal is the Fourier Transform that decomposes periodic signals into a series of harmonic functions of appropriate amplitude. This transform is one of the best methods to study stationary signals because we know the nature of the phenomena to be harmonic a priori. How about when we know nothing of the nature of the signal like the search for techno-signatures? A more powerful tool is requested and a more comprehensive transform is necessary, which is the KLT (see [3]). The KLT has some important advantages:

- it's optimal since it de-correlates the coefficients (are statistically independent)
- it packs the most energy (variance) in the fewest first terms
- it minimizes the MSE (reconstructed compared to original)

- minimizes the total entropy of the sequence
- very good S/N sensibility ad de-noising.

on the other hand, there's also big disadvantages:

- no fast algorithms exist
- heavy computational task
- no fixed basis a priori but generated each time for each observation.

1.2.2 Eigenvalue Decomposition

Several main topics of research like pattern recognition, financial statistics and many more have no request about responsiveness, but in other fields like medical, radar and so on, the time to have the result of a computation is needed almost in real-time. This is the case of radio astronomy where a natural phenomena has a very short impulse. A fast Fourier Transform algorithm exists, but not for the KLT: this is the main issue why it is not so widely used. Nowadays, computer hardwares, equipped with GPU systems that are dedicated to parallel computing, have solved many complex scientific tasks and the KLT could take benefit of this new architectures. In digital signal processing, often, the more the sampled observation is processed the more the analysis is accurate, however this disagree with the request for real-time. In order to extract more eigenpairs (eigenvalues and eigenvectors) as possible, the "Krylov subspace projection methods" are taken in account: Lanczos and Arnoldi are two well known methods of this type (see [4]). The Arnoldi-algorithm is the selected one and it return as a result of the computations a set of eigenvectors obtained by an iterative projection on a given matrix A. The subspace generated from this projection is called Krilov subspace and the iterations for this method are given by the following equations:

$$AV_k = V_k H_k + f_k e_k^T, \quad V_k^H V_k = I_k, \quad V_k^T f_k = 0. \tag{1.6}$$

where:

- A is the symmetric matrix obtained from the signal sample
- V_k are the projections
- H_k is the representation of the projection of A on $K(A, v_0; k)$

First of all, the algorithm does not change the matrix A, so that the Hermitian structure can be exploited and is stored efficiently one time only; second, it returns a k-rank matrix H, with k the amount of eigenvalues of interest relating to the largest eigenvalues of A.

The method consists of the following steps:

1. build the symmetric matrix
2. run the Arnoldi algorithm iterations
3. compute Givens rotations.

Algorithm 1: Arnoldi first step algorithm

Input: (A,v_0)
Output:(V_k, H_k, f_k) such that $AV_k = V_k H_k + f_k e_k^T$, $V_k^T V_k = I_k$and $V_k^T f_k = 0$
1. $v \leftarrow v_1 / \|v_1\|$
2. $w \leftarrow Av$;
3. $H_1 = (\alpha_1)$; $V_1 = (v_1)$; $f_1 = w - v_1\alpha_1$;
4. for j=1,2,3,...,k-1
4.1 $\beta_j = \|f\|$; $v_{j+1} = f/\beta_j$;
4.2 $V_{j+1} = (V_j, v_{j+1})$; $H_j = \begin{pmatrix} H_j \\ \beta_j e_j^T \end{pmatrix}$
4.3 $z = Av_{j+1}$;
4.4 $h = V_{j+1}^H z$; $H_{j+1} = (H_j, h)$;
4.5 $f_j = z - V_{j+1}h$;
5. end

The computational cost of the Arnoldi implementation is $O(kN\log N)$, where k are the eigenvalues of interest. The procedure returns a set of eigenpairs of the matrix H that are a good approximation of those of matrix A. This iteration returns one eigenvector at a time, and for few eigenvectors/eigenvalues no re-orthogonalization is needed. One can point out that for more eigenvalues a restart is needed: this is under investigation. One other aspect of future work is the structure of the matrix A; until now, the correlation of the radio observation was used so a Toeplitz matrix A was generated. Next studies will focus on the covariance matrix and new advanced thresolding methods for signal detection.

In 2003 at the Medicina radio-station facility, the first software was implemented using an advanced system based on a PowerPPC 7400 Altivec (Motorola). The simulated signals were 1024 samples long and the number of eigenvalues extracted 6 (first raw in Table 1.1). After optimizations and more improvements, in 2016 we could compute a simulated signal of 819,200 samples and 90 eigenvalues.

Finally, exploiting the Nvidia CUDA architecture, it is possible to have the new version, last row in Table 1.1 where results are obtained on a laptop Ubuntu 18.04 (Linux) with CUDA 10.0 running on a Intel® Core™ i7 CPU @ 2.50 GHz. The GPU device is GeForce GTX 850M @ 902 MHz, onboard memory 2 GB DDR3, 640 CUDA cores.

Table 1.1 Performance comparison over years

Year	Nr. samples	Seconds	Nr. eigenvalues
2003	1024	23"	20
2016	819,200	61"	90
2019	819,200	17"	90

References

1. C. Maccone, Telecommunications, KLT and relativity, vol. 1 (IPI Press, Library Softbound Edition, 1994)
2. K. Mardia, J. Kent, J. Bibby, *Multivariate Analysis Probability and Mathematical Statistics* (Academic Press, Cambridge, Massachusetts, 1979)
3. R. Dixon, C.A. Klein, On the detection of unknown signals, in *ASP Conference Series*, vol. 47 (1993)
4. D.C. Sorensen, *Implicitly Restarted Arnoldi/Lanczos Methods for Large Scale Eigenvalue Calculations* (Rice University, Texas, Department of Computational and Applied Mathematics, 1995)

Chapter 2
SETI and Temporal Copernicanism

Amedeo Balbi and Milan M. Ćirković

Abstract An often used assumption in astrobiology and SETI studies is that the average condition of the present universe is typical in time, and that many of the relevant physical quantities can be treated as stationary. The uncritical adoption of such assumption of 'temporal Copernicanism' can lead to wrong conclusions, and should be treated with extreme care, if not discarded altogether.

2.1 Introduction

Copernicanism has been a guiding principle in the development of astronomy and cosmology for the past four centuries. The idea is often summarized with the statement that the Earth does not occupy a special place in the universe. More specifically, the assumption of homogeneity and isotropy has been crucial in the development of the standard cosmological model, and has rightly gained the status of a 'cosmological principle' (henceforth CP). Not only has the large-scale spatial uniformity of the universe allowed for progress in our understanding of the universe, but it also made possible to refer our observations to an 'average' spatial reference system, and to assess the typicality of physical quantities with respect to such well-defined average. In other words, when we consider sufficiently large spatial volumes, we can consider them as representative of the universe. This is not only important in cosmology, but also in astrobiology, since it allows us to assume that, although our particular location is far from average in some respects (its position around a certain kind of star, its local chemical abundance, and so on), it can still be considered 'typical' with respect

A. Balbi (✉)
Dipartimento di Fisica, Università di Roma "Tor Vergata",
Via della Ricerca Scientifica, 00133 Roma, Italy
e-mail: balbia@roma2.infn.it

M. M. Ćirković
Astronomical Observatory of Belgrade, Volgina 7, 11000 Belgrade, Serbia

Faculty of Philosophy, Future of Humanity Institute, University of Oxford, Suite 8, Littlegate House, 16/17 St Ebbe's Street, Oxford OX1 1PT, UK

S. Montebugnoli and A. Melis (eds.), *The Search for ExtraTerrestrial Intelligence*, Springer Proceedings in Physics 260, https://doi.org/10.1007/978-3-030-63806-1_2

to some specified reference class (for example, the class of earth-like planets in the habitable zone of their star).

In [9] we argued that a similar appeal to typicality is not possible with respect to time, and, in fact, typicality in time cannot be a well-defined concept in the standard cosmological model. Here we summarize our reasoning and the implications for the search for life in the universe, and in particular for SETI.

2.2 Is Temporal Copernicanism Well-Defined?

The now falsified classical steady-state theory [6] was based on a perfect cosmological principle (henceforth PCP) that postulated homogeneity in both space and time, thus being a special case of CP with the highest level of symmetry. Average in time was a well-defined operation, similar to average in space: any epoch was equal to any other. With the advent of the hot big bang model, PCP was clearly disproved. The now accepted standard cosmological model has a strong evolutionary component (see, e.g., [2, 7]), meaning that the physical average state of the universe has changed radically in time. This has long been recognized as one of the main motivations for astrobiology (see e.g. [12]): while in a steady state universe there would have been an infinite time to produce life, the universe described by the big bang model cannot have been equally habitable at all epochs in the past (and will probably not be in the far future). Thus, there is only a limited window in time for abiogenesis, and the fact that there is at least one planet with life (the Earth) requires an explanation that fits into the available temporal interval. Life requires, for example, the buildup of heavy elements in stars, and possibly even of organic prebiotic molecules in interstellar space and in protoplanetary discs. Typical observers should thus find themselves living in an epoch when the age of the universe is of the same order of magnitude of the lifetime of a main sequence star—a coincidence of time scales that would be unexplainable in a steady-state model [19]. Increasing attention has been devoted, in recent times, to the issue of when the universe started to be habitable and how long it will remain so in the future [14, 15], as well as to the existence of optimal locations and epochs within our galaxy where complex life could preferentially evolve [17].

The evolutionary framework introduced by the big bang model implies that we cannot regard at different epochs as representative of some 'average' state, and that the idea of being 'typical' in time has to be discarded. In short, there are at least three ways that make temporal Copernicanism an ill-defined concept (we refer to [9] for further details). First, some of the objects (e.g. stars and planets, and more generally any bound object) existing in the present universe have not existed in the past and will not exist in the future [1]. Second, the value of any physical quantity fluctuates over long time-scales. Take, for example, a simple (and, at present, well-defined) astrophysical quantity such as the star formation rate (SFR). In what sense can we define an average SFR in time? There is no way of doing this over arbitrarily long scales (for example, the rate would be exactly 0 if averaged over the entire history of the universe, including the future), but only on narrowly specified intervals.

Finally, and more relevant for SETI, when considering the long-term future history of the universe, the problem of intentional influence from advanced technological civilizations (from local effects such as the atmospheric composition of planets, up to mega-engineering on planetary or galactic scale) cannot be excluded: this introduces a completely new kind of uncertainty, that can possibly dominate the values of parameters we wish to average over.

Despite the fact that temporal averages cannot be well-defined in the standard cosmological model, temporal Copernicanism has often been applied to infer general conclusions based on the present state of the universe. One egregious example is the "Doomsday Argument" [13], which attempts to estimate the future duration of humanity from an explicit Copernican premise. A more recent application, relevant to astrobiology and SETI, is the proposal that a dangerous early universe increases the number of habitable sites in the universe today [18]. If temporal Copernicanism cannot be justified, such arguments are clearly not well-grounded.

2.3 SETI and Temporal Copernicanism

A further example, directly relevant for astrobiology and SETI, is the temporal distribution of habitable planets in the Milky Way. This was first calculated in a pioneering study [16] and subsequently elaborated more precisely by various authors [5, 17, 20]. Such studies established that the formation of Earth-like planets started somewhat more than 9 Gyr ago, and has a temporal distribution of ages reaching a maximum (at some billion years ago), then declining to the present-day value. This distribution will conceivably extend for an indefinite amount of time into the future, although with an ever-decreasing rate. Also, as we already pointed out, various studies have addressed the problem of the overall habitability of the universe in time [10, 15]. This kind of investigation is still in its infancy and its conclusions are very uncertain. However, there are no compelling reasons to expect that the probability for the appearance of life in the universe is uniform in time, and in fact there are quite more arguments to the contrary. This very fact clashes with the presumption that we are observing a typical epoch of cosmic history.

The unwarranted assumption of uniformity in time has a direct impact on the chances of success of SETI (or, more broadly, the search for signatures of technological species, or "technosignatures"). In fact, most past studies in the field adopted, explicitly or not, an underlying presumption of stationarity for the appearance of life over cosmic history: one notable example of this approach is the use of the Drake equation [11] to estimate the number of communicating species present in the universe, in which any time dependence of the relevant astrophysical or biological factors is neglected [8]. The shortcomings of such approach are particularly evident when we consider the strict requirement that any electromagnetic signal (or, more generally, any physical interaction) has to satisfy in order for it to be observed by us today. Because of the short crossing-time of our galaxy ($\sim 10^5$ years), the time interval we can directly observe is very small, compared to the age of the universe, for locations within the Milky Way. For example, one can think of an electromagnetic signal of some sort emitted from a location a thousand light-years from the Sun: if such communication ceased before a thousand years ago, we cannot observe it any more. This implies that any technological civilization of which we might find empirical evidence must either be very long-lived or almost coeval to ours [3, 4].

It also implies that the chances of discovering intelligent life outside Earth depend on how life was distributed in time over the course of cosmic history. If we are latecomers, and most other civilizations went already extinct, we may now be alone. Otherwise, if life only started appearing very recently, and the universe is waking up right now, we may be just one of the many examples of infant technological civilizations.

Relaxing the assumption of our temporal typicality can also impact interpretations of the so-called Fermi paradox, i.e. the apparently puzzling fact that we have not encountered any evidence of advanced civilizations in the universe. If the universe was not as life-friendly in the past as it is today, then the puzzle can dissolve, or at least be mitigated. In general, evaluating the outcomes of SETI searches, or even interpreting negative results, needs to assume a knowledge of how the propensity for life of our universe changed over time, a perspective that is inseparable from the historical and evolutionary aspects of the cosmological scenario.

2.4 Conclusions

We have outlined a few reasons that make the definition of average temporal conditions problematic in cosmology and astrobiology. Such problems should suggest the adoption of a more modest, deflationary account of temporal Copernicanism: we cannot really say whether we are living in a typical cosmological epoch and we should therefore refrain from assuming our typicality in this respect and deriving any conclusion from such an assumption. This, of course, does not mean abandoning Copernicanism in general, or, even worse, adopting a 'chronocentric' view of the universe, revolving around the present epoch. Typicality in time can still be well-defined by focusing on subsets of the history of the universe where we might expect a reasonably uniform behavior of some variable quantity relevant to the problem under investigation. For example, when discussing the chances that life evolved on other planets, one might restrict the time interval only to the period when star formation rate is around its peak value (the so-called 'stelliferous' era). But this is precisely the main caveat introduced in [9], that we summarized here: there is no straightforward way in which our typicality in time can be assumed, without accompanying the assumption with some knowledge of the temporal evolution of key factors involved in the habitability of the universe and in the appearance of life. So, at the very least, further work would be necessary to understand the complicated variation in time of the conditions that make the presence of observers possible.

References

1. F.C. Adams, G. Laughlin, A dying universe: the long-term fate and evolution of astrophysical objects. Rev. Mod. Phys. **69**, 337–372 (1997)
2. A. Balbi, Cosmology and time, in *EPJ Web of Conferences* **58** (2013)
3. A. Balbi, The impact of the temporal distribution of communicating civilizations on their detectability. Astrobiology **18**(1), 54–58 (2018)
4. A. Balbi, The spatiotemporal aspects of SETI. Memorie Della Società Astronomica Italiana **89**, 425 (2018)
5. P. Behroozi, M.S. Peeples, On the history and future of cosmic planet formation. Monthly Notices R. Astronom. Soc. **454**, 1811–1817 (2015)
6. H. Bondi, T. Gold, The steady-state theory of the expanding universe. Monthly Notices R. Astronom. Soc. **108**, 252–270 (1948)
7. E. Chaisson, *Cosmic Evolution: The Rise of Complexity in Nature* (Harvard University Press, 2001)
8. M.M. Ćirković, The temporal aspect of the Drake equation and SETI. Astrobiology **4**, 225 (2004)
9. M.M. Ćirković, A. Balbi, Copernicanism and the typicality in time. Int. J. Astrobiol. **19**(2), 101–9 (2020)
10. P. Dayal, M. Ward, C. Cockell, The habitability of the Universe through 13 billion years of cosmic time, arXiv:1606.09224 (2016)
11. F. Drake, The radio search for intelligent extraterrestrial life, in *Current Aspects of Exobiology*, ed. by G. Mamikunian, M.H. Briggs (Oxford University Press, 1965), pp. 323–345
12. S. Dick, *The Biological Universe* (Cambridge University Press, 1996)
13. J.R. Gott, Implications of the Copernican principle for our future prospects. Nature **363**, 315–319 (1993)
14. A. Loeb, The habitable epoch of the early universe. Int. J. Astrobiol. **13**(4), 337–339 (2014)
15. A. Loeb, On the habitability of our universe, arXiv:1606.08926 [astro-ph.CO] (2016)
16. C.H. Lineweaver, An estimate of the age distribution of terrestrial planets in the universe: quantifying metallicity as a selection effect. Icarus **151**, 307–313 (2001)
17. C. Lineweaver, Y. Fenner, B. Gibson, The galactic habitable zone and the age distribution of complex life in the milky way. Science **303**(5654), 59–62 (2004)
18. S.J. Olson, Life-hostile conditions in the early universe can increase the present-day odds of observing extragalactic life, https://arxiv.org/abs/1704.04125 (2017)
19. M. Rees, Cosmological significance of e2/Gm2 and related "large numbers". Comments Astrophys. Space Phys. **4**, 179 (1972)
20. E. Zackrisson, P. Calissendorff, J. González, A. Benson, A. Johansen, M. Janson, Terrestrial planets across space and time. Astrophys. J. **833**, 214 (2016)

Chapter 3
The Newborn European Astrobiology Institute, a Big Opportunity for Outreach and Education

Caterina Boccato and Julie Nekola Nováková

Abstract The European Astrobiology Institute (EAI) is a consortium of European research and higher education institutions and organisations as well as other stakeholders aiming to carry out research, training, outreach and dissemination activities in astrobiology in a comprehensive and coordinated manner, and thereby securing a leading role for the European Research Area in the field. EAI has six scientific Thematic Working Groups and seven activity Working Groups, among which "Outreach, media and corporate identity" is the one we're going to talk in this article together with the "Education" Working Group.

3.1 The Context

3.1.1 General Features of the European Astrobiology Institute

Fundamental questions in science like "How and when did life emerge on Earth?", "How did our solar system and life evolve and how will it develop in the future?" and "Is there life on other celestial bodies?" will not be answered by one discipline alone, but require a concerted and coordinated approach involving many researchers with seemingly unrelated scientific backgrounds.

Also, the European research landscape is rapidly changing on a global scale. Boundaries between disciplines disappear and new cross-disciplinary fields emerge. Astrobiology is one of them. Research in such fields requires interaction and

C. Boccato (✉) · J. N. Nováková
European Astrobiology Institute, La Palma, Spain
e-mail: caterina.boccato@inaf.it

C. Boccato
Istituto Nazionale Di Astrofisica, INAF, Roma, Italy

J. N. Nováková
Charles University, Prague, Czech Republic

S. Montebugnoli and A. Melis (eds.), *The Search for ExtraTerrestrial Intelligence*, Springer Proceedings in Physics 260, https://doi.org/10.1007/978-3-030-63806-1_3

exchange of ideas and new results between scientists from many countries and disciplines, a task that only larger research communities, like the European Research Area, can accomplish.

In order to take astrobiology-related research forward and to prevent a counter-productive fragmentation of the European astrobiology research community through duplicate or excessively overlapping initiatives and structures, the AstroMap Report (drawn up under the EU FP7 programme) unequivocally recommends the creation of a **pan-European platform for research, training outreach and dissemination in Astrobiology**. EAI has been founded in May 2019 during its first General Assembly in Liblice (Czech Republic).

The European Astrobiology Institute (EAI) aims to function as such an entity. Such an institute is required to maintain Europe's leading position in this interdisciplinary field, compared to other countries and regions. The EAI closely collaborate with several related European organisations, including the European Space Agency (ESA) and the European Astrobiology Network Association (EANA), but act as a network of institutions that fundamentally differs from existing bodies [1].

A consortium of representatives of European Research Organisations, which was formed as a result of the initiatives of the COST Action "Origins and Evolution of Life on Earth and in the universe" (Action Identity TD1308), EANA and the Erasmus + Strategic Partnership "European Astrobiology Campus" (EAC), has taken the initiative to create a virtual institute named the "European Astrobiology Institute" (EAI) with the ambition of enabling Europe to emerge as a key player in astrobiology and to develop a general spirit of cooperation and collaboration throughout the European planetary science community. In this way, these communities continue to keep the momentum of the COST Action and EAC initiatives whose grant periods terminated during the Academic Year 2017/2018 and which received excellent reviews (both initiatives were highlighted as success stories by the EU). The *Istituto Nazionale di Astrofisica* to which the first author belongs is one of the six EAI's core organizations.

General aim of the EAI. The European Astrobiology Institute (EAI) is a consortium of European research and higher education institutions and organisations as well as other stakeholders and aims to carry out research, training, outreach and dissemination activities in astrobiology in a comprehensive and coordinated manner, thereby securing a leading role of the European Research Area in the field.

Among the main objectives of EAI, besides performing and disseminating ground-breaking research on key scientific questions in astrobiology with a cooperative interdisciplinary approach, there is also the coordination of public and education activities of European astrobiologists to the general public, industry and all other relevant stakeholders [1].

Public engagement and Education. EAI has six scientific Thematic Working Groups and seven Activity Working Groups, among which there are the two of our interest in this article: "Outreach, media and corporate identity" and "Education". There are also smaller groups on special themes or missions, which have been called

'Project teams', among which the one titled "Science fiction as a tool for astrobiology outreach and education" has already realized an interesting product for public engagement, a sci-fi anthology described below.

3.2 The Outreach, Media and Corporate Identity WG

3.2.1 Corporate Identity

Let me begin with the more specific task of this working group: to establish and manage the corporate identity, and let me open a parenthesis on what really corporate identity is.

A strong corporate identity helps reinforce the Institute's brand image and supports its funds' requests and public outreach activities which, in turn, have impact on corporate identity itself.

When a corporate identity program is presented consistently, it creates a positive and lasting impression of the Institute.

The value of corporate identity is immeasurable: its value increases each time it is presented properly and decreases with every improper application. So what exactly is corporate identity? In short, it's the physical look of your brand. It generally includes a logo and supporting devices, such as your letterhead, business cards and website, all assembled within a set of guidelines. The guidelines govern how the identity is applied and confirm approved designs for printed pieces, color palettes, typefaces, page layouts and all cross media applications that maintain visual continuity and brand recognition. Corporate identity standards ensure that everyone within your Institute is representing the company the same way and every time they interact with all the stakeholders such as peers, large public or, very important, with financial entities.

The Benefit of Corporate Identity. Corporate identity benefits are many and diverse. An Institute that invests in a solid corporate identity tells people "we are here to stay." It's a sign of longevity, which is not only attractive to peers and public, but also is desirable to potential investors. It's an indicator that you are serious about being successful, that you're a reliable leader in your final goals. A functional corporate identity conveys your Institute's ideals, motives and objectives—a sense of what your business is all about. The advantage of creating a consistent and functional corporate identity is that it ensures EAI will be recognized, remembered and respected.

The aims of a corporate identity are to create a single, steadfast and clear visual identity for your Institutes, to project your Institute as a professional, reliable and contemporary organization and to leverage your brand equity and standardize your Institute's visual presentation in all cross media applications.

The Role of Your Logo. At the heart of corporate identity is the logo. As one of your most valuable assets, it should be the first thing your public (large public, funders or peers) sees and the last thing they remember. A well-designed and consistently used logo can unite its members—individual and/or organizations—under one umbrella.

By providing a uniform symbol wherever visual identification occurs, the logo projects quality and professionalism. Your logo should appear on all cross media, including letterhead, business cards, envelopes, checks, marketing materials, advertisements, the web site and more.

EAI Logo Competition. In order to have an effective logo, we launched a competition: in this way we gave space to and stimulated a creative process.

We received dozens of proposals, all the Working Groups' chairs and deputy chairs voted and the winner is a graphics professional Khashayar Teymoori, Art Director of Tunnel Graphic Design Studio in Iran (www.tunnelgraphic.com) (Figs. 3.1 and 3.2).

Fig. 3.1 EAI official logo, extended version

Fig. 3.2 EAI official logo, compact version

EAI Website. We already have, thanks to Wolf Geppert, the EAI chair, a complete website that, now, is going to be re-built according to the new graphics. We're going to discuss its content architecture and the winner of logo context will develop the graphics part of the work. In this second year of life of the Institute, we need a new graphic coherent with the visual identity and at the same time we need also to update and expand the content. Ideally, we should set up an editorial team to take care of updates once something changes, a new project is added, etc.

Once the website is redesigned, up and running, we should create and add more resources for media, a section for the public, and a section for teachers.

We have created a team, with the functions of an editorial board, for the definition of the content architecture, which will follow also the work in progress. We will have the new version of the website by the end of 2020.

EAI Newsletter and Social. The aim of the planned newsletter is to deliver news regarding astrobiology research, opportunities and related news especially to members of the scientific community—not excluding interested members of the public, of course, but it is primarily for the professionals, while the public should engage more with the social media. These should serve both for the professionals (scientists, educators, reporters…) and interested people from the general public. We're working on these tools suffered delays due to the COVID-19 emergency.

For the moment we have the FB page www.facebook.com/EuropeanAstrobiologyInstitute/ and the Twitter account https://twitter.com/EAIastrobiology, both with as a constantly growing number of followers.

Media—Press. Right now, we established a collaboration with press offices of involved institutions. We need to build a database of media contacts for scientists (journalists, science bloggers, TV reporters, etc.). They have also to be sorted by language, field and interests in a such a way that scientists can use properly for sending out their press releases. On the other hand we're going to do a database of astrobiology scientists to be contacted by media (national, European) that will be also sorted by field of expertise and language (nationality).

3.3 The Education WG and Sci-Fi Project Team

The Outreach, Media and Corporate Identity Working Group has to strictly cooperate with Education WG, whose leader is Riho Motlep from the University of Tartu in Estonia, and with the project team "Science fiction as a tool for astrobiology outreach and education". Project teams are smaller groups on special themes or missions that can be led and joined by members of non-participating entities.

In this context we're developing an Astrobiology Bingo Game for the EAI next General Assembly and we have produced a sci-fi anthology with a list of educational links cured by Julie Nekola Nováková.

3.3.1 Astrobiology Bingo Game

The Astrobiology Bingo (or Tombola) is a game of chance in which several people at time can play. It is played with 50 numbered balls and, at least, one card per person (European version) (usually the numbered balls are 90, but in our case we use 50).

The purpose of the game is to mark the numbers on the card (or cards) of each player as they show the balls and becoming the first to complete the whole card (called Tombola).

There are also intermediate wins when a player has 3, 4 and 5 numbers in a row in his/her card, these intermediate wins are called Terna (triad), Quaterna (quadruplet) and Cinquina (five-number row).

In our Tombola anytime a number, with its corresponding image, is taken out of the basket, and then the corresponding image is uncovered, we comment and describe that image. We also stimulate the public to participate to the game telling us what they know about the object represented in the extracted image. This Tombola need at least one presenter, better two presenters. It can be repeated several times.

The Tombola game is made with the use of a big screen with the Tombola's display board where there are 50 boxes covered, these boxes will be uncovered anytime the corresponding number is casually extracted by the participants. In each box there is an image which is significant from an astrobiology point of view (a biomolecule, an extremophiles, a planet, an instrument…)—we have the software that represents the Tombola's display, it runs in any browser.

3.3.2 Sci-Fi Anthology

Science fiction has and for a long time has had a potential to reach wide public and influence its perception of various matters. It may, therefore, serve as a good material for outreach purposes to get actual science across to the public. In a time when the public interest in science dwindles across multiple regions of the worlds and its perception is often skewed by hasty reporting, such a concept that can grasp the attention of people across a wide spectrum of demographics is most useful. Indeed, science fiction has been used in STEM education e.g. in lectures and books by James Kakalios or Andrew Fraknoi, who even keeps an irregularly updated list of SF stories featuring good astronomy [2].

For outreach and educational purposes at the European Astrobiology Institute (EAI), we have compiled an anthology titled "Strangest of All" [3] that contains eight science fiction reprint stories by respected authors, showcasing the following astrobiology topics: (1) life in a subsurface ocean, (2) life under high pressure, (3) possibilities for exotic life in the Kuiper Belt, (4) exotic photosynthesis, (5) Dyson spheres, (6) the SETI program, (7) the Fermi Paradox, and (8) planetary protection.

A nonfiction follow-up article and classroom tips written by the editor and biologist Julie Nováková, who has experience in science and teaching as well as Sci-Fi writing and publishing, accompany each story. The anthology was made available for free download at the websites of the European Astrobiology Institute. Its publication was featured on multiple science outreach and SF literature platforms in several languages. The book was further popularized by follow-up interviews with the authors, who include well-known scientists and at the same time writers such as University of California's Gregory Benford or NASA's Geoffrey Landis (Fig. 3.3).

The book has been downloaded several thousand times as of the time of submitting the abstract, and has elicited good reviews on Goodreads. In the future, we hope to follow up on "Strangest of All" with a print anthology of SF stories written originally for the project in direct collaboration with expert, each of whom will write the nonfiction follow-up. We are also planning to utilize science fiction with good science and interesting themes in other outreach activities such as contests

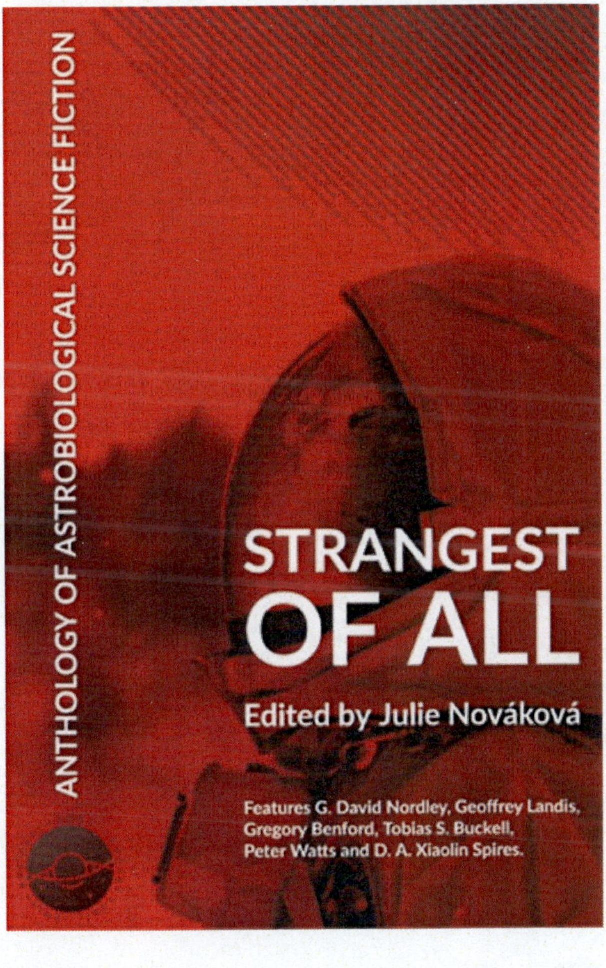

Fig. 3.3 Anthology cover

and exhibitions, and to evaluate the impact of such activities using e.g. Europlanet's Outreach Evaluation Toolkit [4].

References

1. EAI Hompage (2019), https://europeanastrobiology.eu/. Accessed on 30 June 2020
2. Astronomical Society of the Pacific: Education & Outreach: Resource Guides: Science Fiction Stories with Good Astronomy & Physics: A Topical Index. Available at: https://astrosociety.org/education-outreach/resource-guides/science-fiction-stories-with-good-astronomy-physics-a-topical-index.html. Accessed on 28 June 2020
3. J. Nováková (European Astrobiology Institute), *Strangest of All: An Anthology of Astrobiological SF*. Available at: https://www.julienovakova.com/wp-content/uploads/2020/05/Strangest-of-All-Anthology-of-Astrobiological-SF.pdf. Accessed on 28 June 2020
4. Europlanet Society, *Europlanet Evaluation Toolkit*. Available at https://www.europlanet-society.org/outreach/europlanet-evaluation-toolkit/. Accessed on 28 June 2020

Chapter 4
Cherenkov Telescopes for Optical SETI

Patrizia A. Caraveo

Abstract Using Cherenkov telescopes to search for optical signals from interstellar civilizations is a clever way to exploit the vast Cherenkov mirrors and their ultrafast cameras. Such searches could be done in parallel with the telescope routine operations without hampering the gamma-ray data collection.

4.1 A Visionary Common Start

High-energy gamma-ray astrophysics and the search for Interstellar communications have common roots since they have been proposed over 60 years ago by Philip Morrison and Giovanni Cocconi, two visionary scientists who dared to explore new fields.

4.1.1 The Beginning of High-Energy Astrophysics

At the time, cosmic rays represented the frontier of particle physics since studying cosmic ray interactions unveiled new particles. Cosmic rays were studied both directly, through devices capable of tracing their interactions such as stacks of photographic plates brought at high altitude or bubble-chambers on ground laboratories, and indirectly through the air showers they created upon their interactions with the atmosphere' atoms [1]. During such cascade process, the number of secondary particles, carrying a fraction of the incoming particle energy, is rapidly increasing until the multiplication process stops and the shower dies out. Indeed, only the showers produced by the highest energy particles can reach the ground where the secondary particles can be detected using arrays of counters. Although the cosmic nature of the particles' flux was accepted, its origin was far from clear as discussed, e.g., by

P. A. Caraveo (✉)
IASF-INAF Via A.Corti 12, 20133 Milan, Italy
e-mail: patrizia.caraveo@inaf.it

S. Montebugnoli and A. Melis (eds.), *The Search for ExtraTerrestrial Intelligence*, Springer Proceedings in Physics 260, https://doi.org/10.1007/978-3-030-63806-1_4

Cocconi [2]. Unfortunately, detecting and studying the spectrum and the composition of the cosmic radiation could not provide clues on its sources since cosmic magnetic fields, known to exist both at the interstellar and intergalactic scale, deviate the particles' paths. To study the origin of the very-high-energy particles, Morrison in a seminal paper [3] proposed to focus on gamma-rays, the highest energy photons of the electromagnetic spectrum which are certainly produced when high-energy cosmic rays interact with interstellar matter and travel unaffected by interstellar and intergalactic magnetic fields. Moreover, the same processes responsible for the acceleration of cosmic rays could also yield high-energy gamma-rays which could be used to pinpoint their enigmatic sources. Since high-energy photons are absorbed by the atmosphere, Morrison proposed to use balloon-borne detectors, the state of the art technology at the time, prior to the advent of space instruments. At the same time, Cocconi, already familiar with particle induced air showers proposed to exploit the same technique to measure TeV gamma-rays from the Crab Nebula. Indeed, while Cocconi proposed to use an array of particle detectors to measure very high-energy photons able to produce secondary particles energetic enough to reach the ground, others were investigating methods to see the vast majority of air showers that stop in the atmosphere.

In fact, when entering our atmosphere, particles as well as gamma-ray photons interact with its nuclei and produce a shower of highly energetic secondary particles moving with a speed greater than that of light in air (even if their velocity is lower than the speed of light in vacuo). In 1934, the Russian physicist Pavel Cherenkov, working at a particle accelerator, noted that this phenomenon produced bluish luminescence, conceptually similar to the sonic boom in air. It is a very short-lived emission (few billionths of a second) and a very faint one (less than 1/10,000 of the *average* night sky background, but, at nanosec time-scale, the flash exceeds the night sky background). Since high-energy photons and charged particle interact with the atmosphere in a similar way, one must find ways to discriminate showers induced by photons from showers induced by charged particles. Luckily, showers originating from hadrons can be discriminated from electromagnetic ones using their shape since gamma-ray induced cascades are much narrower. The dominant multiplication processes in electromagnetic showers is electron/positron bremsstrahlung, producing gamma rays and electron–positron pairs from gamma-ray conversion. Hadronic showers are wider owing to more complex cascading process involving also pions and muons which render the cascade less symmetric. Moreover, the flashes of Cherenkov light provide a distinctive signature that makes it possible to indirectly reconstruct the energy and arrival direction of the incoming very high-energy photons.

The very first Cherenkov instrument was built in 1953 by Galbraith and Jelley. It was a search-light mirror viewed by a photomultiplier installed in a garbage can which provided protection against stray light [4]. Such pioneering set-up (and those which followed with bigger mirrors and more photomultipliers) detected fast flashes of radiation, showing that the technique was a promising one, but failed to achieve significant results. Indeed, the coming of age of the Cherenkov technique required decades of hardware and software development, as described in [5].

Although both scientists were over optimistic in guessing the flux of potential high-energy sources, their vision somehow started the effort to build gamma-ray detectors and a new branch of astronomy.

4.1.2 Searching for Interstellar Communications

In parallel to their seminal papers on the foundation of high-energy astrophysics, Cocconi and Morrison in 1959 published an equally important paper [6] where they discuss the best strategy to search for a signal from an alien civilization. Decades before the discovery of the first extrasolar planet, these two visionary scientists reviewed the knowledge of the time and concluded that the most promising wavelengths to use is the 21 cm hydrogen line which is not absorbed by the interstellar matter nor by our own atmosphere. To maximize the chances of a detection they suggest that the search should be directed towards nearby stars similar to our sun. Although, very honestly, they admit that it is very hard to quantify the chances of finding a signal, their conclusion is unescapable: "*if we never search, the chance of success is zero*" [6].

This paper inspired Frank Drake to start the first targeted search of a 21 cm radio signal from nearby stars from the Green Bank radio observatory and this later evolved into the SETI program.

Meanwhile the discovery of the laser prompted Schwartz and Townes [7] to re-evaluate the potential of searching interstellar signals at other wavelengths. In a review paper [8] written two decades after [7], Townes concludes that "*the infrared is as good as, and may be a more favorable region for SETI than, the microwave region... However, it does not indicate that we should search only in the infrared or even search at all in this wavelength region with present technology. There is considerable uncertainty as to what design parameters would be considered most critical for interstellar communication by an extraterrestrial civilization. ...But I believe the above discussion does show that we have no assurance the microwave region is the one of choice for a civilization trying to communicate with us. This may affect the scale and style with which SETI is carried out on Earth even in the immediate future*".

4.2 The Current Situation

After the common start, SETI and gamma-ray astronomy proceeded independently for more than half a century. While SETI, after the enthusiastic beginning, faced difficulties, now partly solved owing to the Breakthrough initiatives [9], gamma-ray astronomy blossomed first in space and more recently on the ground [5] through a number of successful projects based on the simultaneous use of several Cherenkov Telescopes.

Currently, three Cherenkov observatories are collecting data, namely:

H.E.S.S. (High-Energy Stereoscopic System), in Namibia (1800 m asl), encompassing 4 telescopes of 12 m diameter arranged to form a square with a 28 m diameter telescope at the center;
MAGIC (Major Atmospheric Gamma-ray Imaging Cherenkov), in La Palma (2225 m asl), characterized by two 17 m diameter telescopes;
VERITAS (Very Energetic Radiation Imaging Telescope Array System), in Arizona (1400 m asl), encompassing 4 telescopes of 12 m diameter.

Although the optical quality of the Cherenkov telescopes is not comparable to that of the optical ones, their big collecting area coupled with large field of view and ultrafast camera have fostered the idea to use them to perform searches for fast optical signals from nearby stars. Such optical flashes from a point-like source should light a single pixel which should appear in the same position in the images collected simultaneously by all telescopes, as opposed to the characteristically elongated pattern due to a gamma-ray induced shower. The possibility to perform the search for fast optical signals without interfering with the normal data collection is certainly a bonus making it possible to monitor targets which happen to be in the telescope field of view.

Thus, after half a century from their common start, gamma-ray astronomy and the search for interstellar communications crossed their paths.

The technique to use Cherenkov Telescopes to search for extraterrestrial signals, described e.g. in [10], was applied by VERITAS to study the peculiar planetary system KIC 8,462,852 [11] which had received much attention since its puzzling behavior prompted explanations based on improbable alien scenarios. The VERITAS team analyzed archival data collected over several observing runs from 2009 to 2015 when the star happened to be serendipitously in the telescopes' field of view. Atmospheric (possibly meteoric) flashes as well as flashes moving through the field of view, probably due to satellite reflections, were seen, but nothing was detected in the direction of KIC 8,462,852. However, the resulting upper limit of 1 photon/m^2 is by far the lowest achieved so far in the search for optical flashes showing the potentialities of Cherenkov telescopes to detect very short optical variabilities during normal operations. Thus, independently from the optical SETI searches, this capability could contribute to the growing field of time domain astronomy.

4.3 A Promising Future

The sensitivity to rapid optical flashes will significantly improve in the future owing to the construction of the most ambitious Cherenkov Observatory known as CTA (Cherenkov Telescope Array) whose baseline design encompasses more than one hundred telescopes divided into two observatories, one in the Northern Hemisphere and one in the Southern Hemisphere, so as to cover the whole celestial sphere [12].

Compared with the current Cherenkov observatories, which rely on few (2–5) telescopes, CTA represents an impressive leap-forward and will be one of the pillars of future astronomy. Each observatory will be made up of different types of telescopes, designed to detect photons of different energies.

Less energetic photons produce a relatively small and faint cone of light. To detect them very large telescopes, positioned close to each other are needed. More energetic photons instead produce a more brilliant burst of light which can be detected by smaller telescopes spread on a large area. Four large-sized telescopes (23 m diameter) will be placed at the centre of the array, surrounded by a dozen medium-sized telescopes (12 m diameter). In the Southern Hemisphere there will also be many dozens of small-sized telescopes (4 m diameter) which will be spread out over a very large area.

All in all, with an enormous collecting area, very big field of view, sensitive detectors and powerful computers, CTA will bring the search for optical flashes to an unprecedented level which will certainly open new windows for optical SETI as well as for time domain astrophysics.

References

1. G. Cocconi, Results and problems concerning the extensive air showers. Rev. Mod. Phys. **21**, 26 (1949)
2. G. Cocconi, On the origin of cosmic radiation. Phys. Rev. **83**, 1193–1195 (1951)
3. P. Morrison, On gamma-ray astronomy. Il Nuovo Cimento **7**, 858–865 (1958)
4. W. Galbraith, J.V. Jelley, Light pulses from the night sky associated with cosmic rays. Nature **171**, 349–350 (1953)
5. P.A. Caraveo, The golden age of high-energy gamma-ray astronomy: the Cherenkov telescope array in the multimessenger era. La Rivista Del Nuovo Cimento **43**, 281–318 (2020)
6. G. Cocconi, P. Morrison, Searching for interstellar communications nature, vol. 184 (1959), pp. 844–846
7. R.N. Schwartz, C.H. Townes, Interstellar and interplanetary communications by optical masers. Nature **190**, 205–208 (1961)
8. C.H. Townes, At what wavelengths should we search for signals from extraterrestrial intelligence? Proc. Nat. Acad. Sci. **89**, 1147–1151 (1983)
9. https://breakthroughinitiatives.org/
10. D. Eichler, G. Beskin, Optical search for extraterrestrial intelligence with air Cerenkov telescopes. Astrobiology **1**, 489–493 (2001)
11. A.U. Abeysekara et al., (VERITAS collaboration) a search for brief optical flashes associated with SETI target KIC 8462852. ApJ. **818**, L33-38 (2016)
12. B.S. Acharya et al., CTA consortium. Intro. CTA Concept Astroparticle Phys. **43**, 3–18 (2013)

Chapter 5
Organics on the Rocks: A Cosmic Origin for the Seeds of Life

Cesare Cecchi-Pestellini

Abstract Interstellar gas phase chemistry is effective in producing a range of simple molecules, including many organic molecules. Such a chemical complexity can be achieved through the processing of mixed ices on the surfaces of dust grains. These complex organics are of great interest to astrobiology, but are simpler than the molecules involved in biological processes. When ice-coated dust grains aggregate together in protoplanetary discs, and eventually in planetesimals, a large volume fraction remains unoccupied. The products of ice processing are retained within these cavities and subjected the repeated processing and additions of metals from the underlying grains. The nature of the chemistry in these cavities is in principle similar to the famous Miller-Urey experiment in which a variety of amino acids was formed. Perhaps clusters of dust grains could be the mechanism of forming and transporting essential molecules to planet Earth.

5.1 Dust Thou Art

Few years ago in 2016, Kepler satellite observations evidenced unusual fluctuations in the light from a star named KIC 8462852. This otherwise normal F-type star is slightly larger and hotter than the Sun, and is located about 1480 light years from Earth, in the constellation Cygnus. The odd behaviour of KIC 8462852, more popularly called Tabby's Star for its discoverer Tabetha Boyajian [1], whose flux dims by a tremendous amount without any regularly repeating signals, brought some researchers to speculate about the existence of alien megastructures. Such a device, named a *Dyson sphere* after the physicist Freeman Dyson who first explored this idea as a thought experiment in 1960, would be a highly advanced piece of technology able to intercept the power of the star, and divert it back to a planet. In an article published in the journal Science [2], Dyson conjectured the existence of stellar power collection systems, with sizes comparable to planetary orbits, that advanced civilizations in our galaxy could

C. Cecchi-Pestellini (✉)
INAF - Osservatorio Astronomico di Palermo, Sicily, Italy
e-mail: cesare.cecchipestellini@inaf.it

S. Montebugnoli and A. Melis (eds.), *The Search for ExtraTerrestrial Intelligence*, Springer Proceedings in Physics 260, https://doi.org/10.1007/978-3-030-63806-1_5

have constructed. Dyson went further and proposed to search for evidences of such structures, as a proof of the existence of such civilizations elsewhere in the Galaxy.

The presence of a Dyson sphere is an extremely fascinating idea, but Tabby's star is not the place where we should look for. A huge number of follow-up observations performed in an extended spectral range, from blue light all the way to infrared light, show that radiation of shorter wavelength is preferentially blocked in all the observed obscuration events. The natural conclusion is that this wavelength-dependent extinction is quite the same as in the general interstellar medium, and it is caused by simple, familiar dust, drifting between us and the star in tendrils of varying thickness. Although the source of the dust remains a mystery (see e.g., [3]), the evidence for aliens around one of the weirdest stars in our galaxy is not looking promising. Should we be disappointed?

Perhaps not. Dust is a crucial ingredient in a remarkable story that links interstellar chemistry with star and planet formation to suggest that the origins of life on Earth—and perhaps elsewhere—may lie in space. Dust grains have a number of absolutely fundamental roles to play in the Universe as we know it today. We may emphasize different parts of the story (for example, the roles of dust in chemistry or in star formation) but it is the comprehensive picture of dust throughout the whole saga that makes it compelling. This complete story makes clear that for a Universe like ours, dust is required to make molecular hydrogen and initiate chemistry, including complex chemistry, and to permit the formation of stars. Dust also provides the raw material for planets, and may even seed them with the molecular building blocks of life.

The interstellar medium is rich in molecules, and especially in organic molecules. It's only in the last half-century that we've begun to realize that the interstellar medium can be not only molecular, but that interstellar chemistry is remarkably complex for what seems a quite hostile environment. Of course, the chemistry of life is much more complex than interstellar chemistry. But wherever life begins, it must start from the kinds of molecules that are readily available in interstellar space. Because the formation of stars and planets is a violent event, the intricate chemical history of the gas from which a planet forms may be obliterated, requiring chemical evolution to be continuously restarted. Nevertheless, the chemical mechanisms that generate biomolecules in space are replicated in protoplanetary systems, so that there is a potential connection between prebiotic organic chemistry and the chemistry we observe on going in the interstellar medium. Nature has had a four-billion-year head start in implementing controlled chemical evolution, and in doing that a central role has been played by dust.

5.2 Making Complex Molecules from Simple Interstellar Ices

The inventory of the molecular Universe is continually progressing [4]. The nature of the selection of molecular species is not transparent. Hydrogen and helium atoms make up the bulk of all atoms in the Universe. Of the remaining 0.1%, half comprises oxygen atoms, and a quarter is carbon. All the other elements are compressed into the left 0.025%. It is surprising then, if we look at the list of the approximately 200 molecules discovered in space, carbon-containing molecules dominate, making organic chemistry the norm. In regions that are almost unimaginably cold and dark, molecules are slowly forming, bit by bit, over literally millions of years. The key feature of the chemistry going on in space is that it takes a long time. But still this is not enough. Interstellar space is too cold for most chemical reactions to occur, as the low temperature makes it hard for molecules drifting through space to acquire the energy needed to break their bonds, and form new others to make larger molecules. However, occasionally, the gas molecules collide with the dust grains. Some freeze there to form molecular ices, and exposed to the harsh space environment can be transformed by surface and bulk interactions into more complex molecules, before being released again into the gas.

Astronomers have known for decades that there is a substantial amount of water in space, and in fact water ice is the most abundant interstellar solid material. As a dominant form of oxygen, the most abundant element in the Universe after hydrogen and helium, water controls the chemistry of many other species. Water molecules in the interstellar gas were detected very early in the history of astrochemistry. In 1969, Charles Townes and colleagues detected strong water emission in a line with a wavelength of 1.38 cm, in the radio part of the spectrum [5]. Interstellar water ice, as distinct from gaseous water molecules, was first identified by Gillett and Forrest in 1973 [6], and is now widely detected in interstellar dark clouds.

Dust grains in cold astronomical regions build up appreciable ice mantles through the accretion and subsequent surface chemistry of atoms and molecules from the gas. The timescale for an atom or molecule to collide with a grain and stick to it (the so-called *freeze-out time*) is inversely proportional to the molecular number density; in dense clouds it is of the order of 100,000 years or lower, generally smaller than cloud lifetimes. Despite the fact hydrogen is a very weakly bound species, hydrogen atoms have a residence time on dust grains at temperatures of 10–20 K long enough to react with oxygen to form water molecules. The ice formed is not pure, but it may contain CO, CO_2, and other less abundant species. Such species may stick to a grain directly from the gas-phase, or be formed in the same way as water, with carbon combining with hydrogen to make methane, or nitrogen with hydrogen to produce ammonia. These molecules are frozen over by a layer of ice, preserving them for over millions of years. Such a favourable circumstance enables a dramatic chemical transformation to take place in the interstellar space. Ices containing such simple species can be "cooked" in various ways to produce more complex species. Such a processing can be driven by fast charged particles, energetic radiation, in particular ultraviolet photons and X-rays, as well as through the warming up of the ice. The energy injection provides the activation of molecules initially present in the ices, allowing the development of an important chemical reactivity.

What kind of chemical complexity can we expect? Chemistry in interstellar environments is in many ways very different from the chemistry in terrestrial chemical laboratories that we are more accustomed to. Evidently, basic concepts remain the same, but the physical conditions are such that some of the usual hypotheses are no longer valid. The exotic reaction methodologies offered to synthetic organic chemists to forge new chemical bonds are of course not available in space, and the synthesis must occur through processes that a modern chemist undertaking routine synthesis in a laboratory would consider insufferable or archaic. The basic idea is that radicals trapped in the iced mantles acquire mobility and react forming complex organic molecules as the dust temperature reaches $\sim$30 K [7]. The origin and quantity of the trapped radicals is uncertain; they could be e.g., pieces of iced species broken by the ultraviolet photons or, the results of the incomplete hydrogenation of the simple parent species. Generally, their nature should reflect the ambient physical and chemical conditions, and the radiation environment.

The complex species arising into the ice may be released to the gas-phase, where they are eventually observed, through a variety of processes. Thermal desorption seems a possible cause; it becomes sufficiently rapid in interstellar conditions when the ice temperature rises to about 100 K, but is totally negligible for typical ice temperatures of 10 K. Species on or near the surface of a dust grain can be desorbed by photons of the interstellar radiation field. Where extinction is high so that the interstellar radiation field plays no significant role, photodesorption may be driven by the weak radiation field created when cosmic rays ionize hydrogen atoms and molecules. The subsequent recombination spectrum of ions and electrons generates an ultraviolet field [8]. As noted first by Léger and collaborators in 1985 [9] the passage of heavy cosmic rays through dust grains deposits heat in the grains. On large grains, the heat is deposited in a cylindrical volume along the track of the cosmic ray, and thermal desorption of weakly bound molecules such as CO can occur from the heated areas of the surface at the intersection of the cylinder and the surface. Where the grains are small enough, the cylinder encompasses the entire grain, so desorption occurs from the entire surface area of small grains. Other mechanisms involve desorption driven by surface chemistry, grain explosions, and grain-grain impacts. The efficiency of a process typically depends on the involved molecular species, and it is mediated by the energy transfer at the molecular level (see e.g., [10]).

The organic molecules discovered in space until few years ago consisted of a backbone of carbon atoms arranged in a single and roughly straight chain. In 2014, astronomers announced to have for the first time detected towards the galactic centre—in the Sgr B2(N) molecular cloud—a carbon-containing molecule called iso-propyl cyanide, in which the carbon structure branches off in a separate strand [11]. Two years later, another complex organic molecule, propylene oxide was found again in Sgr B2(N) [12]. It is a chiral molecule, meaning that it exists in non-superimposable forms that are mirror images of one another. Both of the characteristic traits of these molecules are essential features of terrestrial biochemistry, and reflect the evidence that organic matter can be naturally synthesized in space. The discovery that such kind of molecules exist well outside our Solar System, long before a planetary surface is created, suggests that the potentially prebiotic chemistry traced by asteroids and comets in our planetary system could be replicated elsewhere in the cosmos, supplying with these life-bearing elements newly born planets.

Although, thus far, amino acids have not been identified in the interstellar medium, many organic compounds, some of them directly linked to our biochemistry, are routinely produced by cosmic chemistry, and are likely widespread throughout the Universe. Among them, two species with the peptide moiety, isocyanic acid (HNCO) and formamide ($HCONH_2$) were detected [13, 14] just three years after radio astronomy opened the way to the molecular Universe, with the discover of ammonia in 1968 [15]. This cannot be considered a fortuitous event, and in fact, these pretty simple molecules are incredibly abundant in our Universe. Their detections together with other complex organic species would have an added value if they occur in the gas surrounding a young solar-type star, implying that their synthesis would be coeval with the formation of planets in a protoplanetary system.

5.3 Is the Origin of Life Linked to Cosmic Chemistry?

The formation of a star is a violent and chaotic process in which the gas is flowing in and ejected outwards at speeds up to hundreds of kilometres per second, as the gravitational infall is locally opposed by thermal, turbulent, and magnetic pressures, by dynamical outflows, and—since the parent cloud is rotating—by angular momentum effects. As a consequence of all such competing process a cloud contracting to form a solar-type star forms a swirling disc. Initially, discs rapidly funnel material onto the star but, as the surrounding molecular core is used up or otherwise disperses, the accretion rate decreases, and only a small amount of the original material persists in the disc. Such discs can be considered protoplanetary not only for the geometry of the Solar System, but also for the high detection rate of exoplanets.

The physical conditions in a protoplanetary disc vary greatly, with hot and dense regions of gas and dust near to the star and much colder material at greater distances from it, providing a suitable background in which cosmic chemistry may be replicated. The Atacama Large Millimetre Array (ALMA) has allowed a detailed observation of molecules in these regions. While CO, CO_2, HCO, and H_2CO are often abundant species in the cold zones of the disc, CH_3OH or CH_3CN are only found in a few regions, and more complex organic molecules are not observed [16–18]. Such evidence is in striking contrast with the analysis of the debris of the planetary construction in the Solar System, where quantitative analyses confirm the existence of a wide variety of organic species inside certain types of meteorites [19], and to a lesser extent in comets [20]. In principle, these compounds can form during the assembly of a protoplanetary system. The chemical inventory includes alcohols, amines, amides, esters, amino acids, the building blocks of proteins, amphiphiles, the building blocks of membranes, simple sugars such ribose, a crucial piece of the chemical machinery inside cells, and nucleobases, the building blocks of RNA and DNA. Some of those amino acids [21] and sugar derivatives [22] contain significant excesses of enantiomers having the same handedness of terrestrial biomolecules. Why is such a wealth of complex organics not detected in the gas-phase of discs?

The number of complex species observed in a FU Orionis object (V883 Ori, [23]), with abundances comparable to those in comet 67P/Churyumov-Gerasimenko [20] demonstrates that ice chemistry in discs does proceeds similarly as in hot-cores. FU Orionis events can be described as an abrupt mass transfer from the accretion disc onto a young, low mass T-Tauri star. Thermal instabilities initiate an outburst. The duration of the outburst is determined by viscosity within this hot, ionized area. The rise time of these eruptions is typically of the order of 1 year, but can be much longer. This high-accretion, high-luminosity phase has typical lifetimes of decades. Such a sudden increases in the luminosity of the central star quickly expands the snow lines into the disc, creating a "sublimation front". The natural conclusion is that normal condition, desorption mechanisms fail in supplying complex species to the gas.

All these considerations suggest that icy mantles onto dust grains are the repository of chemical complexity in the early stages of a disc lifetime. While their ice mantles are chemically evolving, dust grains grow through collisional agglomeration. This process ends up in the formation of loosely packed structures with much of dust aggregate internal volume being vacuum and trapped ices. Interstitial voids must occur even in highly organized, densely packed structures. For example, dense packing of spheres in a face-centred cubic lattice leaves 26% of space unoccupied. In 2000, Walt Duley pointed out that in the interiors of dust aggregates, those internal voids produced by the accretion process offers the possibility of a radical change in the chemistry [24], i.e. the re-accretion of reaction products onto other components of the aggregate. As desorbed products can be in an energetic state, these secondary reactions might mimic some aspects of high-temperature chemistry. Dust aggregates can also be impulsively heated by cosmic ray impacts, during which cosmic rays suffer negligible energy losses. During the interaction, sputtering may also occur, with atoms of the grain materials (such as Si, Ca, Mg, Fe, and P) dislodged and ejected from an inner or outer surface. The heat released during the collision may lead to the partial or total vaporization of the ice filling the cavities. As a consequence, part of the chemical species forming the ice enter a transient, warm, high-pressure gas phase, together with sputtered atoms from the grain substrate, in a hydrogen-rich atmosphere. Thus, the cavities within the grain aggregate bring together all the components of gas and dust, a unique situation outside planetary systems. The resulting scenario is a reasonable analogue of the conditions that Stanley Miller envisaged as plausible for the primitive Earth atmosphere in its famous experiment at the University of Chicago [25]. Therefore, grain aggregates may represent the equivalent in space of terrestrial micro-laboratories containing raw materials of reducing chemical composition suitable for conversion into complex organic species. The final products are likely to be very similar to those obtained from laboratory chemistry under terrestrial conditions. Because of the reducing atmosphere in the cavities large organic molecules are allowed to form, with the inclusion of sputtered atoms from the grain substrate.

5.4 Where Do We Go from Here?

A common perception of the chemistry in space environments is the tendency toward uncontrollable reaction pathways that produce complex mixtures containing fundamental prebiotic building blocks and a plethora of undesirable and intractable products. The fact that a number of molecular components of contemporary organisms can be produced in the laboratory does not necessarily mean that they were essential for the origin of life, or that they were prebiotically available. In space there are many mechanisms by which biochemically interesting species can be synthesized. Not all pathways are equally efficient, but the broad range of conditions under which organic compounds form demonstrates that their synthesis can occur in a wide variety of astronomical regions. We do not have a coherent scenario for the origins of life on Earth. What we do know is that many organic compounds, including several of those crucial to the functioning of modern organisms, are routinely produced by cosmic chemistry, and are likely widespread throughout the Universe. What remains problematic is the understanding of how such chemical entities may have spontaneously organized in systems that can self-assemble, process information, orchestrate reaction pathways, and ultimately self-replicate.

If the starting point for our story is straightforward, the subsequent steps are uncertain, made up of things we know and understand with speculations founded on intuition, together with some ideas based on imagination rather than science. It's a complicated situation, but our journey in search of the lost roots of life has certainly begun, is clearly heading in the right direction, and is still in progress.

References

1. T.S. Boyajian, D.M. LaCourse, S.A. Rappaport, et al., Planet Hunters IX. KIC 8462852—where's the flux? MNRAS 4, 3988–4004 (2016). https://doi.org/10.1093/mnras/stw218
2. F.J. Dyson, Search for artificial stellar sources of infrared radiation. Science **131**, 1667–1668 (1960). https://doi.org/10.1126/science.131.3414.1667
3. P. Woods, Melting moons make Tabby's star murky. Nat. Astron. **3**, 961–961 (2019). https://doi.org/10.1038/s41550-019-0946-1
4. B.A. McGuire, 2018 census of interstellar, circumstellar, extragalactic, protoplanetary disk, and exoplanetary molecules. ApJ Supp. Ser. **239**, 17 (2018). https://doi.org/10.3847/1538-4365/aae5d2
5. A.C. Cheung, D.M. Rank, C.H. Townes, D.D. Thornton, W.J. Welch, Detection of water in interstellar regions by its microwave radiation. Nature **221**, 626–628 (1969). https://doi.org/10.1038/221626a0
6. F.C. Gillett, W.J. Forrest, Spectra of the Becklin-Neugebauer point source and the Kleinmann-Low nebula from 2.8 to 13.5 microns. ApJ **179**, 483 (1973). https://doi.org/10.1086/151888
7. R.T. Garrod, E. Herbst, Formation of methyl formate and other organic species in the warm-up phase of hot molecular cores. A&A **457**, 927–936 (2006). https://doi.org/10.1051/0004-6361:20065560
8. C. Cecchi-Pestellini, S. Aiello, Cosmic ray induced photons in dense interstellar clouds. MNRAS **258**, 125–133 (1992). https://doi.org/10.1093/mnras/258.1.125
9. A. Legér, M. Jura, A. Omont, Desorption from interstellar grains. A&A **144**, 147–160 (1985)

10. M.R.S. McCoustra, J.D. Thrower, Exciton-promoted desorption from solid water surfaces, in K. Wandelt (ed.) *Encyclopedia of Interfacial Chemistry* (Elsevier, 2018), pp. 383–395. https://doi.org/10.1016/B978-0-12-409547-2.13910-1
11. A. Belloche, R.T. Garrod, H.S.P. Müller, K.M. Menten, Detection of a branched alkyl molecule in the interstellar medium: iso-propyl cyanide. Science **345**, 1584–1587 (2014). https://doi.org/10.1126/science.1256678
12. B.A. McGuire, P.B. Carroll, R.A. Loomis, I.A. Finneran, P.R. Jewell, A.J. Remijan, G.A. Blake, Discovery of the interstellar chiral molecule propylene oxide (CH_3CHCH_2O). Science **352**, 1449–1452 (2016). https://doi.org/10.1126/science.aae0328
13. L.E. Snyder, D. Buhl, Interstellar isocyanic acid. ApJ **177**, 619 (1972). https://doi.org/10.1086/151739
14. R.H. Rubin, G.W.J. Swenson, R.C. Benson, H.L. Tigelaar, W.H. Flygare, Microwave detection of interstellar formamide. ApJ Lett. **169**, L39 (1971). https://doi.org/10.1086/180810
15. A.C. Cheung, D.M. Rank, C.H. Townes, D.D. Thornton, W.J. Welch, Detection of NH_3 Molecules in the interstellar medium by their microwave emission. Phys. Rev. Lett. **21**, 1701–1705 (1968). https://doi.org/10.1103/PhysRevLett.21.1701
16. K.I. Öberg, V.V. Guzmn, K. Furuya, C. Qi, Y. Aikawa, S.M. Andrews, R.A. Loomis, D.J. Wilner, The comet-like composition of a protoplanetary disk as revealed by complex cyanides. Nature **520**, 198–201 (2015). https://doi.org/10.1038/nature14276
17. C. Walsh, R.A. Loomis, K.I. Öberg et al., First detection of gas-phase methanol in a protoplanetary disk. ApJ Lett. **823**, L10 (2016). https://doi.org/10.3847/2041-8205/823/1/L10
18. L. Podio, F. Bacciotti, D. Fedele et al., Organic molecules in the protoplanetary disk of DG Tauri revealed by ALMA. A&A **623**, L6 (2019). https://doi.org/10.1051/0004-6361/201834475
19. A.S. Burton, J.C. Stern, J.E. Elsila, D.P. Glavin, J.P. Dworkin, Understanding prebiotic chemistry through the analysis of extraterrestrial amino acids and nucleobases in meteorites. Chem. Soc. Rev. **41**, 5459–5472 (2012). https://doi.org/10.1039/c2cs35109a
20. L. Le Roy, K. Altwegg, H. Balsiger et al., Inventory of the volatiles on comet 67P/Churyumov-Gerasimenko from Rosetta/ROSINA. A&A **583**, A1 (2015). https://doi.org/10.1051/0004-6361/201526450
21. J.R. Cronin, S. Pizzarello, Enantiomeric excesses in meteoritic amino acids. Science **275**, 951–955 (1997). https://doi.org/10.1126/science.275.5302.951
22. C. Cooper, A.C. Rios, Enantiomer excesses of rare and common sugar derivatives in carbonaceous meteorites. Proc. Natl. Acad. Sci. U.S.A. **113**, E3322–E3331 (2016). https://doi.org/10.1073/pnas.1603030113
23. J.-E. Lee, S. Lee, G. Baek, et al., The ice composition in the disk around V883 Ori revealed by its stellar outburst. Nat. Astron. **3**, 314–319 (2019). https://doi.org/10.1038/s41550-018-0680-0
24. W.W. Duley, Chemistry in grain aggregates: a source of complex molecules? MNRAS **319**, 791–796 (2000). https://doi.org/10.1046/j.1365-8711.2000.03820.x
25. S.L. Miller, A production of amino acids under possible primitive earth conditions. Science **117**, 528–529 (1953). https://doi.org/10.1126/science.117.3046.528

Chapter 6
COGITO in Space

Daniela de Paulis

Abstract *COGITO in Space* is an experiential narrative sending thoughts into outer space as radio waves. The project exists both as a mobile installation and as performative event staged inside the cabin of the Dwingeloo radio telescope in The Netherlands. For both versions of the project, a team composed by three neuroscientists prepare the subject with a lab grade electroencephalogram (EEG) device and a virtual reality (VR) headset, showing an experimental video of the Earth seen from space. The brain activity stimulated by the video is recorded and simultaneously transmitted into space in real time, using the antenna of the Dwingeloo radio telescope.

6.1 Introduction

The first brain activity was purposefully sent into space in 1977, etched as sound in the Golden Record by Carl Sagan and his team. The brain activity of Ann Druyan was recorded for one hour on June 3 1977, with the help of Julius Korein and Tim Ferris of the New York University Medical as a single electroencephalography (EEG) channel and compressed into a one-minute sound recording [2] (see Fig. 6.1).

Although a highly poetic gesture, the brain activity etched in the Golden Record does not convey actual information about human mental states or the human existential condition. Neuroscience has progressed much since, with modern EEG recording and analysis techniques allowing sufficient fidelity to capture individual brain states to the point where they can be used as individual 'fingerprints', and for the decoding of subjective and cognitive processes.

COGITO in Space began as a thought experiment in 2013, with the first public talk about the project as part of KOSMICA in Mexico City in August 2013. A public performance and radio transmission of brain activity into space followed as part of

This paper was first published on the #46 issue of "Antennae: Journal of Nature in Visual Culture" and Leonardo MIT Journal 54, Issue 1 [1].

D. de Paulis (✉)
IAA SETI Permanent Committee, Roma, Italy
e-mail: selavyrose@gmail.com

S. Montebugnoli and A. Melis (eds.), *The Search for ExtraTerrestrial Intelligence*, Springer Proceedings in Physics 260, https://doi.org/10.1007/978-3-030-63806-1_6

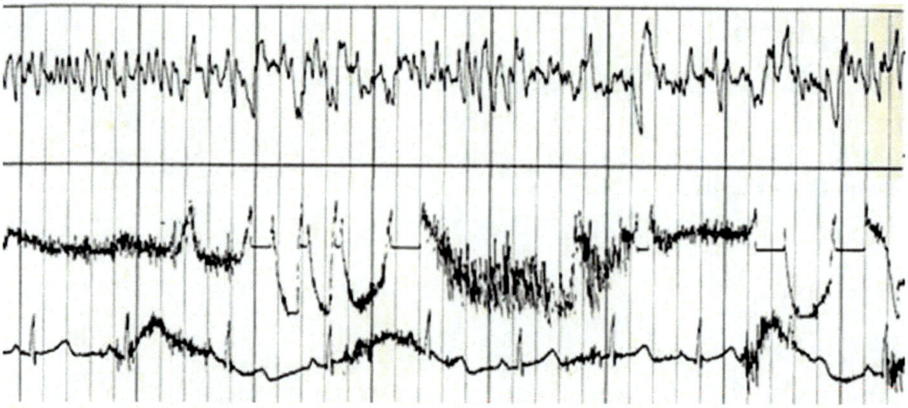

Fig. 6.1 A sampling of Ann Druyan's brain waves, recorded on June 3, 1977 (*Credits* NASA)

the 50th Design Biennale in Ljubljana in September 2014. Over the years before its completion, the project has been widely presented as work in progress at conferences, festivals, art events, gradually becoming more concrete at each presentation, thus starting to take shape in the mind of the public longer before its staging at the Dwingeloo radio telescope in November 2018. The first public presentation of the project as a live performance was at the GOGBOT new media art festival in Enschede in The Netherlands, in September 2017, followed by another live performance in February 2018 as part of the TEC ART international festival in Rotterdam. During this event, the brain activity of participants was recorded and transmitted into space as part of a live audio-visual performance, in remote connection with the cabin of the Dwingeloo radio telescope and in collaboration with radio operators of the CAMRAS team.

6.2 Conceptual Background of the Project

COGITO in Space is an interdisciplinary, collaborative project and the result of six years research at the Dwingeloo radio telescope and ASTRON, the Dutch Institute for Radio Astronomy. The intellectual background of the project grew over the years, thanks to the contribution of specialists from different fields, however my reasoning behind it focuses mainly on two concepts: the unresolved question of Mind and Body Dualism and the *Overview Effect*. Another important reference was the novel *Solaris* by Stanislaw Lem and the homonymous film by Andrei Tarkovsky.

The reasoning around the project began as a reflection upon the use of the electromagnetic spectrum, and especially radio waves, by radio astronomers for gathering data of cosmic phenomena, and upon the type of knowledge we gain from the discovery of such remote events that cannot be known through direct sensory experience. Working alongside radio astronomers over the years, I had the opportunity to better understand the relevance of radio transmissions in contemporary thought and their great cultural impact. My questioning focused especially on how radio waves transmissions have been expanding the human reach into the cosmos, towards places where manned space exploration might never be able to reach, allowing us to remotely explore outer space and exposing the conventional perception of our surroundings to a virtual yet uncannily real and detailed landscape, made of matter still to be defined (see Fig. 6.2). Radio waves have gradually become the mean for the virtual human space travel and the carriers of a new-found cosmic awareness and cultural contents, exceeding their scientific and technological functions. One of the questions that originated from my research and that eventually became one of the leading concepts of *COGITO in Space* is "how does the knowledge acquired through remote observation of the universe influence our cognition and how does the mind interact with the matter of such distant universe and vice versa?" [3].

The photo of the first hole drilled on the Martian surface was pivotal for me in questioning the relation between the body and the mind in contemporary cosmology. The photo portrays a landscape upon which the human intervention has imprinted its mark, showing in great details the nature of the action and the surrounding landscape. The uncannily realistic scene is symbolic in my opinion of what I call the Dualistic Problem in Contemporary Cosmology: the bodily, sensory response to the familiar

Fig. 6.2 The very realistic photo of the first sample drilled on the surface of Mars. The hole called John Klein (on the right) was drilled by NASA's Curiosity rover on Feb. 8, 2013. *Credits* NASA/JPL-Caltech/MSSS [4]

looking matter, texture and colours of the landscape portrayed in images of cosmological objects, is held back by the mind, that reminds us of the abstract nature of the phenomenon represented in the image, belonging to a world that humankind might never experience directly through the senses. In this respect, radio waves allow us to travel with our thoughts much farther and faster than our senses, through a realm of abstract cosmological spaces. As cosmology progresses in the discovery of the universe, the role of the mind in the interpretation of the picture of 'reality' remains largely unknown. The way our mind works might be fundamental to understanding the universe, the mirroring effect between mind and matter might be essential to comprehend our very unique—and possibly arbitrary—notion of the universe and our life within it.

The conceptual framework underpinning *COGITO in Space*, touching upon notions of philosophy of mind and the unresolved debate on the mind–body Dualism, eventually aims at creating a connection between the mind of the viewer and his or her idea of 'universe' and delving into the subjective existential questions on the origin and meaning of life.

The project reflects also upon the paradox of the human mind to overcome its own limitations, pushing the boundaries of knowledge towards what lies beyond itself. A paradox well expressed by philosopher Thomas Nagel: "Isn't it sufficient to try to understand ourselves from within, which is hard enough? Yet the ambition appears to be irresistible, as if we cannot legitimately proceed in life just from the point of view that we naturally occupy in the world, but must encompass ourselves in a larger world view. And to succeed, that larger world view must encompass itself" [5].

I named the project *COGITO in Space* following my reasoning on the concept of Dualism, the title in fact is poetically referring to the changes of the understanding of fundamental concepts in Western thought—such as the concept of human COGITO—towards a contemporary culture increasingly characterized and influenced by space exploration. The title thus suggests the expansion of the concept COGITO as conceived in an era where notions regarding human form and movement were guided by anthropocentrism and framed within the Euclidean space, into the contemporary understanding of relativity and cosmological phenomena. Contemporary physics challenge the anthropocentric view and assert that space and time are not absolutes that extend equally throughout the universe. In the Theory of Relativity in fact, space and time differ according to the movement of the observer. The virtual cosmonaut in *COGITO in Space* starts her journey within the Cartesian mind, localized in the brain and accurately measured through electrical signals, reaching for interstellar space where spatiotemporal references blur and eventually fade into the unknown, alike the radio signal carrying the thoughts.

Since its early development in 2013, *COGITO in Space* has been informed by conversations with space philosopher Frank White, co-founder of the Overview Institute and author of "The Overview Effect, Space Exploration and Human Evolution", an influential essay in which he investigates the cognitive shift happening in the mind of the astronauts after witnessing the sight of the Earth from outer space. Frank White reports that as a result of the *Overview Effect*, the mind of the astronaut develops a notion of a unified planet and of greater ecological and social awareness [6]. This concept deeply influenced the development of *COGITO in Space*. The immersive video I created especially for the project in collaboration with film maker Sandro Bocci, aims at evoking the *Overview Effect* while at the same time leaving space for open interpretation and critique rather than functioning as an outreach platform by endorsing a particular current of thought. The video includes some of the footage captured by the camera onboard the International Space Station, showing the curvature of the Earth as the station rotates. The footage however is greatly reinterpreted, combining the documentarist material with experimental film techniques evoking in the viewer memories, an aesthetic experience as well as sensory feedback.

The images mimicking organic forms, intend to induce flickering, ancestral memories of our evolutionary past, from stellar dust up to our present place in the cosmos, for which the evolutionary history of the universe, life, consciousness is perceived as a narrative intrinsic to one's body and mind. The footage in virtual reality manages to bring the images as close as possible to the retina, engaging the deep mind directly, rather than creating a field of view. Combining experimental footage with realistic views of the Earth seen from space, the video attempts to disrupt the familiar image of our home planet, inducing a more introspective journey into the perception of our place in the cosmos, delving more deeply into subjective memories and mental associations.

Towards the end, the video shows the iconic Blue Marble image, taken during the last manned lunar mission in 1972: supposed to show humankind its place in the cosmos, the photo seems to emphasize instead that we know our home planet only through photos. We don't know our home, we only know it through our subjective interpretation of the image, filtered through our individual memory and thought process. In my opinion, the lack of understanding of our home planet plays an important role in the struggle for the search of our identity as earthlings and our role in the universe. By receiving feedback from some of the project's participants, I realized that the sight of the blue planet, following a series of abstract images, surprises the viewer and inspires in her mind deep emotions of belonging and an almost mystical resonance beyond thought (see Fig. 6.3).

From the start of the project, I was interested in the idea of sending brain waves into space generated by the sight of our home planet, shifting the Earth-centred perspective to the Cosmos-wide mindset. This vision seemed to me evocative of a universal enough existential narrative, for which the human being is compelled to explore the unknown while grieving the consequences of leaving a part of herself behind.

COGITO in Space thus speculates about a hypothetical future in which humans might be confronted with the possibility of leaving planet Earth for the void of outer space. What would be the impact of this extreme journey on the human existential condition? The poetical reference for this aspect of the project is the novel *Solaris* by Stanislaw Lem, which addresses the psychological conflict of losing one's identity and memories of terrestrial life, letting go of one's sense of belonging to Earth during space travel. The experience of the project aims at creating a mirroring effect between the unknowns of the cosmos and inner subjectivity, echoing the thoughts of Dr. Kelvin, *Solaris*'s main character: "Man has gone out to explore other worlds and

Fig. 6.3 COGITO in space VR video still image. Image by Sandro Bocci, 2017. © Daniela de Paulis

other civilizations without having explored his own labyrinth of dark passages and secret chambers, and without finding what lies behind doorways that he himself has sealed" [7].

6.3 *COGITO in Space* in Relation to Previous Interstellar Messages

The broad questioning underpinning the development of *COGITO in Space* led me to reflect upon the complexity of human thought as possibly the most recognizable signature of our experience of life. When faced by the neuroscientists with the decision on which part of the thinking process to highlight in the brain activity recording for the project, I eventually opted for the possibility of representing the entire dynamics of the brain and to capture as much as possible the contradictory and often conflicting facets of the thinking process. As a result, the neuroscientists and I focused on recording spontaneous cognition, also known as 'stream of consciousness', the raw thinking process that is typically not object of study in neuroscience. The term 'stream of consciousness' was coined by philosopher and psychologist William James in "The Principles of Psychology" in 1890: "consciousness, then, does not appear to itself as chopped up in bits … it is nothing joined … it flows. A 'river' or a 'stream' are the metaphors by which it is most naturally described. In talking of it hereafter, let's call it the stream of thought, consciousness, or subjective life. In literary criticism, stream of consciousness is a narrative mode or method that attempts to depict the multitudinous thoughts and feeling which pass through the mind" [8]. The idea of recording and sending into space a 'stream of human consciousness', challenges the notion of interstellar message as conceived by scientists, notoriously Frank Drake and Carl Sagan amongst others, who composed and transmitted into space the Arecibo message in 1974. This and the following interstellar messages, focused on representing humankind's scientific and technological knowledge, emphasizing the mathematical and logical notion of intelligence.

COGITO in Space focuses instead on the continuous thought process that occupies most of our individual narrative, on the existential struggle and controversial nature of a sentient being, on her or his deepest thoughts and doubts, asking a potential extraterrestrial listener "are you as controversial as we are, do you share our inner struggle? Can we empathize on what we don't know and cannot understand about our existence and the meaning of life?"

The two different approaches to communicating with extraterrestrial intelligence are also highlighted in the novel *Solaris.* While hovering on a spaceship over the surface of the sentient planet Solaris, the scientist of the mission prompts Dr. Kelvin, the mission psychologist, to transmit his brain activity towards the planet, while focusing on the scientific purpose of the operation. However, while the electrodes are attached to his head and the transmission of his brain activity takes place, Dr. Kelvin lets his mind wander through memories of his recent and old past, contaminated and overlapped with dreams.

6.4 Code for Interstellar Transmission

The artistic requirement of recording the brain activity of the entire brain and conveying the complexity of human thought proved to be very challenging from the scientific and technological point of view. As an artist I value scientific accuracy, in fact I consider the scientific language as the 'material' of the project, making it an essential requirement that people experiencing *COGITO in Space* would not be deceived regarding the accuracy of their experience, within the limits of current understanding of brain activity analysis. For the project we use a lab grade electroencephalogram (EEG) device (see Fig. 6.4). The EEG research has been developed in collaboration with neuroscientists Robert Oostenveld, Stephen Whitmarsh and Guillaume Dumas.

The EEG reading is able to detect the complex pattern created by the unique interaction of the brain waves in time, showing that the mind is engaged in negotiating different stimuli, produced both by the sensory and psychological experience. The EEG reading however is not invasive and cannot detect the object of thoughts, therefore it presents no privacy concerns.

For *COGITO in Space* the neuroscientists developed a unique code for interstellar transmission, the first of its kind, both accurate from the artistic and scientific perspective. The code manages to reliably convert the 'river of consciousness' measured by 32 electrodes, into a mono sound that can be converted into a radio signal that can be transmitted into space using a powerful antenna.

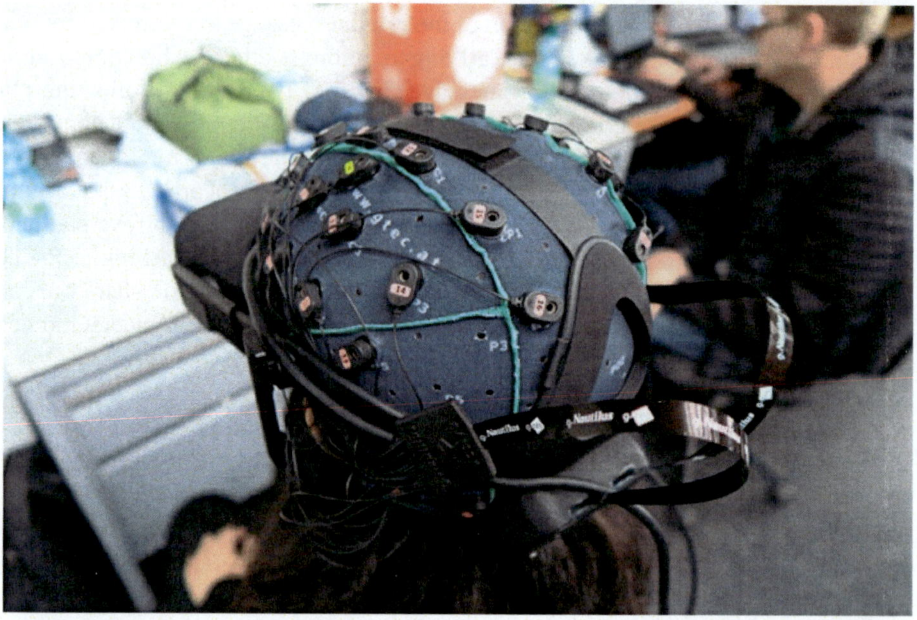

Fig. 6.4 The 32 electrodes EEG device used in COGITO in space. Photo: Sandro Bocci, 2017

Additionally the code manages to retain the complexity of the original signal very accurately, communicating across space the spatiotemporal and psychological nature of our advanced mammalian brain. A potential extraterrestrial intelligence would be able to reconstruct the signal and perhaps understand how it was created and the complex nature of our thinking. However due to the great cosmic distances it is extremely unlikely that the radio signals will ever be detected and interpreted before they fade into the background noise of the universe. The project has been presented at several international SETI (Search for Extraterrestrial Intelligence) conferences and discussed with specialists in the field. As a member of the permanent SETI committee myself, I am very aware of the controversy in transmitting powerful signals into space in the attempt to communicate with an extraterrestrial intelligence. For this reason I chose not to target a particular celestial object with the *COGITO in Space* transmissions, thus keeping the transmitting antenna still and spreading the signal across the sky instead.

The work in progress for developing the first code for interstellar transmissions using EEG signals, was a very interesting part of the work: every two months the entire team met for short research retreats at the ASTRON guesthouse, the living and working facilities adjoining the scientific facilities and designated for radio astronomers. The guesthouse as well as the cabin of the Dwingeloo radio telescope became our spaceship, our safe ground control for working together: radio astronomers, neuroscientists, radio operators, artists, brainstorming ideas, sharing languages and methodologies, having in depth conversations about all aspects of life (see Fig. 6.5).

Fig. 6.5 The COGITO in space team inside the cabin of the Dwingeloo radio telescope. Photo: Sandro Bocci, 2017

6.5 Performance at the Dwingeloo Radio Telescope

On the 5 November 2018, *COGITO in Space* finally settled at the Dwingeloo radio telescope, its originally intended physical location. The project in fact was designed from the start to be presented inside the cabin of a radio telescope that would mimic the post-human outer shell of the body, its life support system in a half human-half technological system. The Dwingeloo radio telescope was inaugurated in 1956 and established as part of the Dutch heritage in 2011. The cabin still features some of the original radio equipment, together with modern amateur radio devices and its environment has become the iconic filmic set for the ongoing performances of *COGITO in Space* since 2014.

I envisioned the event at the radio telescope to be an experience for the visitors rather than a conventional exhibition opening, with several activities taking place throughout the day, in form of cinematic reality for which the extra-ordinary and ordinary mixed. The day started at the ASTRON auditorium with a symposium and keynote lectures by space philosopher Frank White, cultural anthropologist Fred Spier, retired NASA astronaut Nicole Stott and moderated by art critic Josephine Bosma. Frank White introduced the *Overview Effect* and his more recent concept of *Cosma Hypothesis*, suggesting that while exploring outer space, humankind should avoid repeating the mistakes committed during colonial exploration of Earth. Fred Spier is professor of Big History at the University of Amsterdam. Big History is a cutting edge field of research investigating the development of both natural and cultural events, from the Big Bang to the present, working along a very broad time scale from an interdisciplinary perspective, merging human sciences with physics, astronomy and geology. Spier spoke about his experience of the Moon landing as a child and the cultural meaning of the Earth Rise photo, which for many years had a strong impact only on the American narrative, leaving Europe and the rest of the world almost indifferent. Astronaut Nicole Stott spoke about her personal experience of the *Overview Effect* from the International Space Station and her life on board, together with her crew mates.

The symposium was followed by a conversation between myself, the neuroscientists and Josephine Bosma, who asked about my background as a contemporary dancer and how that affected the making of *COGITO in Space*. In my answer it emerged that the physical division between oneself and outer space is a perceptive illusion, as matter at the atomic level is a continuity, the difference between one's body and outer space being an instance of intensity and density of matter: the blood running through my veins as I write is in fact as close to my skin—the body's membrane between interior and exterior space—as the air and materials touching it. My internal organs are thus a continuity with outer space. In *COGITO in Space* the body streams beyond its protective skin into the infinite space beyond it, poetically drifting towards the unknown.

The talks session was followed by a meditative lecture-walk in the area surrounding the scientific facilities, the Dwingeldverveld National Park, mostly a flat landscape dotted with heath lands. The walk, guided by planetary scientist Maarten Roos, took the visitors along a straight line going from the radio telescope into the wide open landscape and back, allowing for the visitors to gaze as far as the horizon, as if viewing the curvature of the Earth. The walk, lasting approximately one hour, was interspersed by cogitations on the origin of the cosmos, of life on Earth, the existence of possible extraterrestrial life, and was informed by Fred Spier's book "Big History and the Future of Humanity".

The walk aimed at inspiring a sense of belonging to Earth before virtually leaving the planet, as well as creating room for individual existential questioning. The walk was followed by an introduction into the history of radio astronomy and the Dwingeloo radio telescope by astronomer Roy Smits. The short presentation led the visitors inside the cabin through a seamless transition between the different events taking place and the different people performing them. While the visitors started entering the cabin and taking their seats, I turned the engine of the radio telescope on, while the participant lying on a gravity chair was being prepared by the neuroscientists with the EEG device and the VR headset. As the people sat still, the radio telescope started tracking the star Betelgeuse, the departing point for the brain activity transmission, and this triggered the rotation of the cabin. Once reached the complete stop of the rotation, the curtains lowered, shortly after the video in virtual reality started playing while the brain activity was recorded and simultaneously transmitted into space, with the antenna of the radio telescope pointing still towards the sky (see Fig. 6.6).

Fig. 6.6 The artist and radio operator transmitting brain waves into space, 5 November 2018. Photo: Sandro Bocci, 2018

This action allowed for the brain activity to be spread across a large portion of the sky with the Earth's rotation. The brain activity recording and the radio transmission happening simultaneously were visualized and projected onto video inside the cabin in real time (see Fig. 6.7). The sound produced by the brain activity created hypnotic and repetitive patterns that generated a meditative mood inside the cabin: people experiencing the event seemed to draw their attention inwards and joined

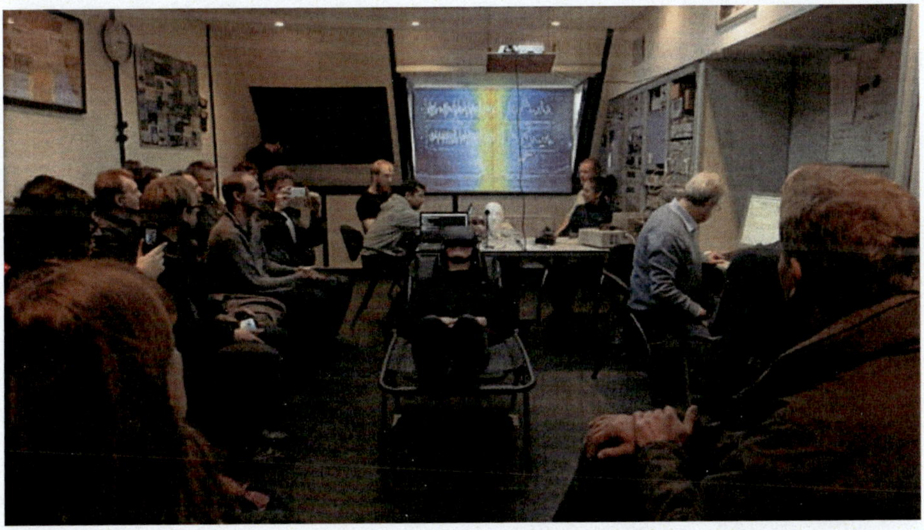

Fig. 6.7 Live brain activity transmission into space. Dwingeloo radio telescope, 5 November 2018. Photo: Sandro Bocci, 2018

the participant in her intimate journey into outer space with the mind. The entire performance lasted approximately forty five minutes and was repeated twice during the day. Following the event on 5 November, participants contacted me reporting to have lived a special day and that the experience will remain in their memory as a unique moment. Their feedback suggested that my intention to convey the project as a subjective experience, as well as a personal sequence of neuro-images, was brought to reality.

The performance intentionally used an objective and abstract language, assimilating the event to a scientific presentation, my intention being to allow for the experience to find its niche in the individual memory according to one's inner understanding, regardless of the specialist aesthetic appreciation or knowledge by the participant and the observer. The work in progress of the project and the event at the Dwingeloo radio telescope have been documented as part of a reportage filmed by Sandro Bocci that is touring internationally at the time of writing.

6.6 Conclusions

Drawing from diverse fields of research and sources of inspiration, *COGITO in Space* is a multilayered art project, whose aim is to generate questions and meanings that might resonate and mirror very differently from person to person. The project is intentionally left dualistic, controversial and open ended and despite its complexity, both conceptual and technological, it can be appreciated by people of all ages and cultures. The written record of individual experiences that is being collected throughout the ongoing presentations of the project, might provide over time some insightful material on the subjective perception of the visualization of Earth from space.

References

1. D. De Paulis, COGITO in space, antennae. J Nat. Vis. Cult. **46**, 199–211 (2019), D. De Paulis, The Metamorphosis of a Periplaneta Americana, Leonardo MIT Journal 54, Issue 1 (2021)
2. T. Phillips, *Voyager, The Love Story* (2011), https://science.nasa.gov/science-news/science-at-nasa/2011/28apr_voyager2. Accessed on 16 Sept 2018
3. D. De Paulis, COGITO, radio astronomy and neuroscience in art. Acoustic Space **1**(16), 177–186 (2017)
4. M. Wall, *Curiosity Rover Drills Into Mars Rock, Collects Sample—A Space First* (2013). https://www.space.com/19708-mars-rover-curiosity-rock-drill-sample.html. Accessed on 18 Sept 2018
5. T. Nagel, *Mind & Cosmos* (Oxford University Press, New York, 2012).
6. F. White, *The Overview Effect* (American Institute of Aeronautics and Astronautics, Reston, 2014).
7. S. Lem, *Solaris* (Faber & Faber, London, 1970).
8. J.A. Cuddon, *A Dictionary of Literary Terms* (Wiley Online Library, 2012).

Chapter 7
SETI Program at the Medicina INAF Radioastronomy Station: Past, Present, Future of PC Based Spectrometers

Roberto Lulli, Giuseppe Pupillo, and Stelio Montebugnoli

Abstract From 2012, the development of a new line of low-cost, high-resolution spectrometers began at the Medicina Radio Astronomical Station following the guidelines in the article "SETI back end made inexpensive" presented at the 61st International Astronautical Congress (IAC) by Dr. Eng. Stelio Montebugnoli in 2010 (Montebugnoli et al. (2010) Proceedings of 61st international astronautical congress 2010 (IAC 2010), IAC-10.A4.4, vol 5, pp 4284–4288).

The growing availability of high-performance:

- Commercial Off-The-Shelf components (COTS), such as multi-core Central Processing Unit (CPU) Personal Computers (PC),
- computational accelerators such as Graphics Processing Units (GPU), Intel Xeon Phi many-core systems,
- Field Programmable Gates Arrays (FPGA),

has allowed the development of a set of instruments whose costs increase linearly but performances increase exponentially. Moreover, the use of ad-hoc software with an open architecture and without extreme optimizations has made these tools scalable and easily integrable even with new and more powerful accelerators and data acquisition boards produced over the years. These spectrometers can be exploited in different radio astronomical amatorial or professional applications. The bandwidth from the narrow ones typical of space debris and NEO (Near Earth Objects such as

R. Lulli (✉) · G. Pupillo
National Institute for Astro-Physics (INAF), Radio Astronomy Institute (IRA), via Piero Gobetti 101, 40129 Bologna, Italy
e-mail: roberto.lulli@unicam.it

R. Lulli
School of Advanced Studies (SAS), Camerino University, via Camillo Lili 55, 62032 Camerino, Italy

S. Montebugnoli
National Institute for Astro-Physics (INAF), Radio Astronomy Institute (IRA), former engineer in charge of the Northern Cross Radio Telescope, Via Fiorentina 3513, Medicina 40059, Italy

National Institute for Astro-Physics (INAF), Scientific Directorate, SETI advisor, Viale del Parco Mellini 84, Rome 00136, Italy

S. Montebugnoli and A. Melis (eds.), *The Search for ExtraTerrestrial Intelligence*, Springer Proceedings in Physics 260, https://doi.org/10.1007/978-3-030-63806-1_7

comet asteroids, etc.) detection and tracking, to larger bands typical for SETI observations. Here following, we report all the developed systems, their performances, the real field tests we carried out till now, and the guidelines for further developments.

7.1 Introduction

7.1.1 PC Based Spectrometers of the 90s at the Medicina Radio Astronomical Station

At the Medicina Radio Astronomical Station, the development of PC based digital spectrometer started in the 1990s as soon as CPU processing capabilities showed a significant increase with the availability on the market of Intel Pentium and Pentium II class processors.

The first projects named Sentinel 1 and Sentinel 2 were spectrometers dedicated to the Radio Frequency Interference (RFI) detecting and monitoring, a problem that at the Medicina Radio Astronomical Station in the 1990s became more significant.

The Sentinel 1 (see Fig. 7.1) is based on a high-end Intel Pentium class commercial PC with a Data Acquisition (DAQ) board connected to the Peripheral Component Interconnect (PCI) bus. The Sentinel 1 DAQ board integrated a Texas Instruments C62 Digital Signal Processing (DSP) microprocessor used to compute the Fast Fourier Transform (FFT) on the input data [2]. In Sentinel 1, the PC's CPU is used to control the entire system and to manage the data visualization and storage. In this configuration, the Sentinel 1 was able to analyze a 10 MHz bandwidth with a maximum of 1024 channels (10 kHz spectral resolution) [3].

The solution to offload some particular data computation to another internally integrated computing system is used whenever the main system CPU doesn't have

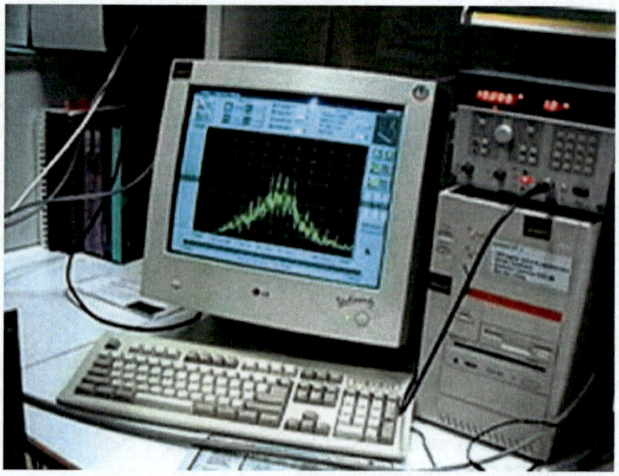

Fig. 7.1 The Sentinel 1 spectrometer installed at the Medicina Radio Astronomical Station (Credits to Stelio Montebugnoli, IRA—Medicina Radio Astronomical Station)

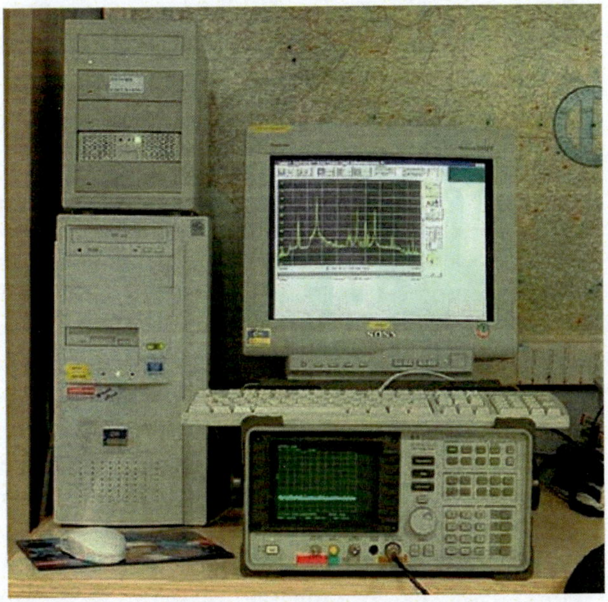

Fig. 7.2 The Sentinel 2 spectrometer installed at the Medicina Radio Astronomical Station (Credits to Stelio Montebugnoli, IRA—Medicina Radio Astronomical Station)

enough computing power for the algorithm or must be left free for other important tasks. This solution is also typical today both on high level scientific and industrial applications and on the consumer market (game console).

The Sentinel 2 (see Fig. 7.2) is based on a high-end dual-socket Symmetric Multi-Processing (SMP) Intel Pentium II Xeon PC [3]. Like the previous one, Sentinel 2 is equipped with a DAQ board. It is connected to the PCI bus that was not supported by an onboard DSP microprocessor CPU because the two Xeon Processor had enough computing power both to calculate the FFT and to control the system, the user interface, and the data storage. In this configuration, the Sentinel 2 was able to analyze a 20 MHz bandwidth with a maximum of 4096 channels (5 kHz spectral resolution).

Both the Sentinel systems proved to be so flexible that they could also be used in other applications such as the EMBLA [4] and SETI projects. As far as the recent research is concerned, Sentinel systems, albeit inadequate due to limitations in computing power that led to low spectral resolution, were an excellent test for the development of future projects.

The Sentinel systems were dismissed during the 2000s because of the rapid obsolescence and reduced temporal reliability of the platform, mainly when used 24 h per day and 365 days per year.

7.1.2 PC Hardware Evolution of the 2000s

In the 2000s, several problems appeared on the CPU market: the development of silicon foundries with higher integration scale required increasingly important investments that only a large market with the high added value could allow to amortize. So the number of companies able to have a state of the art silicon foundry was reduced to a few units.

In that period, it was also understood that the increase in the computing capacity of the microprocessors no longer grew linearly with the clock frequency. Moreover, in 2005, Intel failed to market its latest microprocessor (Pentium 4 at 4 GHz clock frequency) because it was unstable when used with air-cooled heat sinks. The most computing capable Intel microprocessors of that period absorbed over 100 W of power and produced a quantity of heat very near to the package dissipation technical limit. These problems led Intel and other companies to stop the MHz race (which had characterized previous decades and which in recent years had turned into the race for GHz) and to change technology roadmaps.

The solution to the problem was found, considering that the increase factor of the power dissipated by a microprocessor grows approximately with the square of the clock frequency growth factor. Theoretically, it was possible to replace a high clock frequency capacity and significantly reducing energy consumption. This solution introduced the so-called multi-core CPU, a CPU composed of two microprocessors in the same chip at half clock frequency (concerning to the previous one) still maintaining the same order of computing power, but significantly reducing energy consumption.

Although considering the technological limits of Moore's law [5], from the second half of the 2000s, these CPUs have evolved rapidly both in their internal architecture and in the number of cores integrated into a single chip.

The CPU computing power has evolved dramatically even if the clock frequencies of the individual cores have remained in the order of 4 GHz. However, according to Amdahl's law [6], to make the most of the computing capacity of the new CPUs, the application software has to be modified through the integration of algorithms and methods of parallel computing.

In the same period, there was an introduction to the market of the so-called many-core architectures. "Multi-core simply means "more than one core," but when the number of cores grows well beyond the reach of finger counting, we use another name. Many-core chips are multi-cores that contain tens, hundreds, or even thousands of cores. While there is no hard threshold beyond which a multicore becomes a many-core, an easy distinction is that you probably have a many-core if you no longer care about losing one or two cores" [7].

Examples of many-core architectures are Nvidia and AMD GPUs and the Intel Xeon Phi. GPUs were initially being developed to raise the computing power of PC video adapters, but their architecture it also proved suitable for developing parallelizable algorithms typical of DSP applications. In the GPUs, the Control Unit (CU) and the cache management are elementary respect to the one of CPU. A single Control

Unit manages not just a single Arithmetic and Logic Unit (ALU) but tens or hundreds of them all executing the same instruction on different data of the same array. The space saved on the die by using a simple CU was used to implement a high number of ALU. Actual Nvidia GPUs have thousands of core, managed by tens on CUs named Streaming Multiprocessors (SMs) [8].

Intel Xeon Phi system derived from a failed project of a high-performance GPU. They can be considered a hybrid between a multi-core CPU and a GPU because it has similarities with both. Each Xeon Phi integrated into the same chip tens of cores much straightforward than the ones of contemporary Intel CPUs but equipped with computing units dedicated to Single Instruction Multiple Data (SIMD) operations [9].

The use of many-core architectures was boosted with the porting of many standard numerical libraries (such as those used for linear algebra or FFT) by rewriting the original code entirely. The architecture was modified to take advantage of many-core hardware capabilities to make easy and speed-up the porting of old applications on the newly available accelerators. These libraries guarantee a considerable increase in computing performance of some algorithms (especially in those which by their nature are more easily paralleled—e.g., the FFT) with minimal changes to the original code [10, 11].

The FPGAs are digital integrated circuits that are designed to be configured by a customer after manufacturing—hence the term "field-programmable" [12]. They are composed of an array of blocks containing both combinatorial logic and some memory in the form of flip-flops. Each block is connected to others by a network of interconnections that can be changed programming a set of switches. So by programming the combinatorial logic and the flip flops connections of each block together with the set of network switches, it is possible to implement fast and straightforward digital circuits but also complete microprocessors. Many FPGAs can be reprogrammed (reconfigured) each time they are powered on letting to implement each time a different digital application.

FPGAs let the users have a very flexible digital circuit with almost the performance of an Application Specific Integrated Circuit (ASIC) but more comfortable to set up. It is very cheap to implement for mass productions under 10 k units because it doesn't need to deal with the typical problem of silicon foundry projects.

Starting from the 2000s were introduced on the market advanced FPGAs that also included in the same chip CPUs, Static and Dynamic Random Access Memory (SRAM–DRAM), DSP units, and other dedicated and optimized logic blocks.

A typical use of FPGAs is glue logic (a custom chip installed between standard chips with different data exchanging standard to let them communicate) or custom logic for low production high demanding applications. They are widely used in research fields with projects were data rates and processing need the state of the art of technology. In the last years, FPGAs programming was made easier year by year: although programming via functional blocks and Hardware Description Language (HDL) languages (e.g., Verilog and VHDL), it is always possible. Today it is also possible to use high-level languages suitably modified with appropriate libraries and extensions like OpenCL [13, 14]. So many FPGA manufacturers like Intel and Xilinx put on the market high-end FPGAs based accelerator board, but the prices are still today quite high (order of €10,000).

In the years, many-cores and FPGAs market became so crucial that these products shared the same semiconductor lithographic processes, and they were composed of the same number (order) of transistor than high-end CPUs. The price of these products decreased, and the Software Development Kits (SDKs) evolved, letting to commonly use them for various High-Performance Computing (HPC) applications (scientific computing, industrial simulations, Big Data analysis, Artificial Intelligence (AI), etc.).

7.1.3 New Design Guidelines for PC Based Spectrometers

In 2010, Dr. Eng. Stelio Montebugnoli, searching for a new approach to develop a new line of spectrometers suitable for radio observations and following the computing hardware evolution of the 2000s, proposed at the annual IAC in Prague (CZ) a new idea for low-cost spectrometer design.

Dr. Montebugnoli proposed to design a low-cost, high-resolution spectrometer for SETI researches using COTS components, like standard multi-core and/or multi-socket PCs, industry-standard DAQ boards, and GPUs or FPGAs as computing accelerators.

Dr. Montebugnoli concluded that "The algorithm described in this article for SETI investigation can be implemented on several systems with a low and medium cost. Taking into account the advantages and disadvantages shown, the user can find the best solution (in terms of costs, performances, the difficulty of programming, I/O rate, power consumption, processing bandwidth, etc.) to make its SETI back end. Even with a small budget, it is possible to reach high-performance back end, especially using GPU systems" [1].

7.2 The Design of the New Spectrometers

7.2.1 NewSpec: The First Approach for a New Spectrometer

Following the new design guidelines, the first approach to design a spectrometer began in 2012 with the development of the NewSpec project. The base objective of this project was to design and develop a low cost (order of €10,000), high-resolution (Hz order resolution) SETI spectrometer to replace the dismissed Serendip IV [15]. But, the final goal was to have a final spectrometer able to analyze all the 500 MHz Intermediate Frequency (IF) bandwidth coming from the first L band receiver (1.35–1.45 GHz) of the 32 m parabolic dish (Medicina RT-32) without another downconversion.

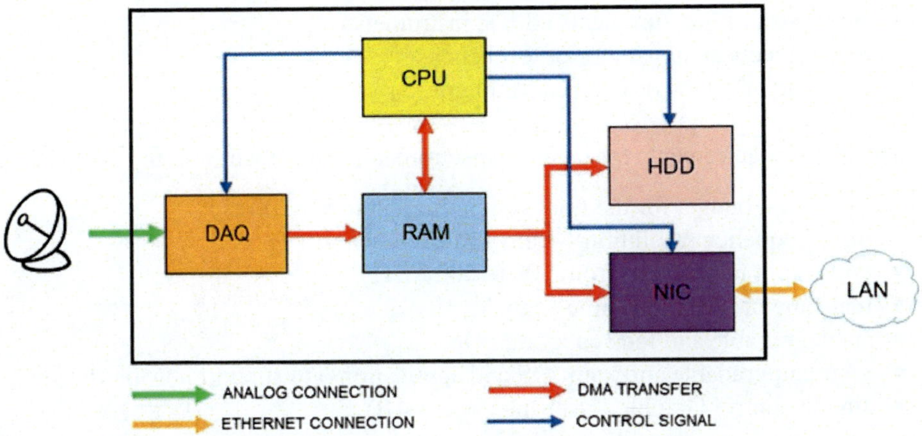

Fig. 7.3 Block diagram of the NewSpec hardware design

The first approach to developing such a spectrometer was to use a system based on a high-performance PC with an Intel 3.6 GHz four-core i7 3820 CPU hosting on the PCI Express bus a 1,5 GS/s, 8-bit resolution DAQ board Ultraview AD8-1500 × 2. All the data computing tasks were carried out by the CPU that also managed Direct Memory Access (DMA) internal data handling, operating system control, and data storage and visualization (see Fig. 7.3). The algorithm used for spectrum analysis was the FFT implemented in the Fastest FFT in the West (FFTW) library [16–18].

In this configuration, the system was able to analyze the entire Medicina RT-32 IF bandwidth with a maximum of 4 million channels at variable resolutions. The spectrometer was also able to reach the Hz resolution, but not in real-time because it required a substantial decimation of the sampled signal due to the impossibility of the acquisition board to work at bandwidth lower than 250 MHz.

The spectrometer had a cost of about €14,000 (a significant part of which was devoted to the acquisition board) and the CPU capabilities limited its performances. Moreover, the system did not have high scalability and reliability mainly because the DAQ board software drivers were available by very few Operating Systems (OS).

The spectrometer operated adequately only for a few months because an OS update was enough to put the system out of order [20].

7.2.2 Spectrometer's Design Final Requirements

After the NewSpec experience, Dr. Montebugnoli carried out another global analysis of the project to refine guidelines and spectrometer final requirements.

The final goal of the project was to develop a low cost, long term, very open system that should be available for:

- SETI applications,

- single sets of measurements ad-hoc instruments,
- new projects demonstrator or test bench,
- test bench for new researches paradigm.

The initial requirements for the new spectrometer project were set to:

- low cost—budget for the final system limited to €10,000;
- high-frequency resolution—from 100 Hz down to 1 Hz;
- wide input bandwidth: from 100 to 500 MHz;
- real-time or quasi-real-time analysis;
- stable, reliable and long supported OS;
- secure upgradable software (OS and applications) to support new hardware;
- simple control Graphics User Interface (GUI).

During the work, these requirements were integrated with other desired features:

- a flexible software architecture to let the implementation of different SETI processing chains;
- an open and scalable software architecture to let the use of the spectrometer in other Medicina Radio Astronomical Station activities involving the FFT processing: e.g., dedispersor in the frequency domain for high sensibility pulsar studies, bistatic radar receiver for NEO and space debris monitoring, beamforming for antennas array;
- possibility to replace the FFT with other spectral analysis techniques like Wavelets, Karhunen-Loève Transform (KLT), Hilbert Huang Transform (HHT), Agnostic Entropy analysis, etc.;
- measures and results encoded in different data formats, both scientific standards and application-defined.

7.2.3 Spectrometer's Hardware Design

The more definitive constraint was the overall unit cost. Spectrometer's cost depends on many factors: performance-related factors like input bandwidth, spectral resolution, timings (e.g., real-time or quasi-real-time analysis), and technical related ones like R&D costs, components chosen, and produced units number. In our requirements, the cost limit was related only to the components.

The €10,000 limit cut off the possibility of developing dedicated ASICs and also the use of high-end FPGA boards. As an example, a single ROACH or ROACH2 board [21, 22], or an FPGA PCI Express accelerator board costs, at that period, exceeded the available budget. Moreover, some DAQ boards equipped with high-end FPGA (like the one used in the NewSpec project) are out of budget.

The solution found was to use a straightforward design, similar to the one of NewSpec, exploiting, as a base, a standard PC with a multi-core CPU. Here were installed a DAQ board with a lower cost than the NewSpec project one, and if needed,

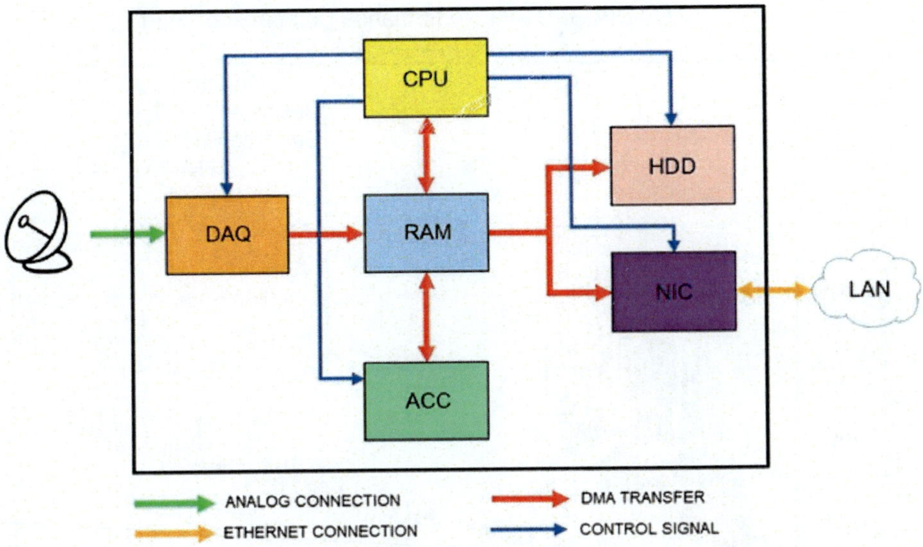

Fig. 7.4 Block diagram of the spectrometer hardware design

raising the computing power by one or more many-core accelerator boards (see Fig. 7.4).

Considering budget limits, as DAQ, we evaluated both internal boards and external adapters typically used in electronics measure applications as fast oscilloscopes, that due to high volume production, have a much affordable price.

We tested many boards, both older ones connected to the PC by PCI or High-Speed Universal Serial Bus (USB) buses and newer ones connected by PCI Express or Super-Speed USB buses. In the tests of these devices, we observed that external adapters were much more resilient against RFI than the internal boards. In a more in-depth test of the latter, made without input signals and with input connectors closed on the characteristic impedance, we observed in the spectra many sharp and narrow lines at specific frequencies related to the PC clock value.

This situation, although not critical in SETI applications because the post-processing algorithms remove these artifacts, leads to problems in other applications where it is advisable to perform an accurate calibration of the instrument before measurement.

About the accelerator boards, we evaluated Nvidia GPUs, AMD GPUs, and Intel Xeon Phi boards. Our performance evaluations were focused on many algorithms but mainly on FFT, and they were carried out using the manufacturer's free available SDKs. The test results on FFT computations showed better performances of Nvidia GPUs with CuFFT software library [23] than the comparable (by price, computing power, and semiconductor technology) products of other manufacturers. Both in FFT and other algorithms test, results were generally similar to others in the scientific literature [24–27]. The latter usually reports Nvidia GPU computation tests on General Purpose GPU boards, namely Tesla product line. These products had (and also today

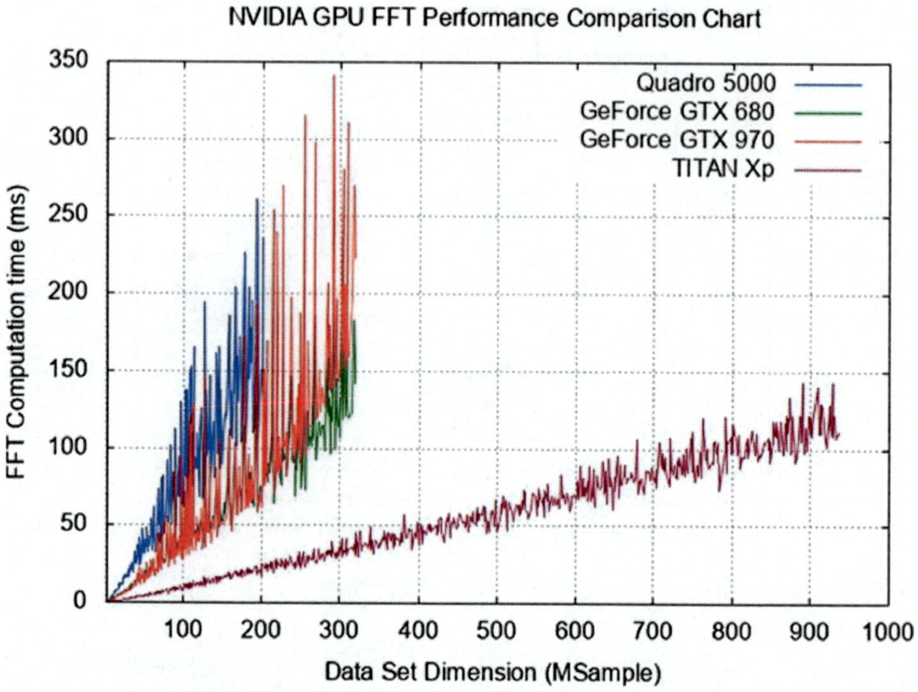

Fig. 7.5 Comparison chart of the performance in FFT computation time of different Nvidia GPU boards with increasing data set dimensions (one test every million of points)

have) an high cost that exceeded our limited budget, so we were forced to substitute Tesla boards with high-end products of the entertainment dedicated GeForce and Titan line.

A summary chart of the most significant Nvidia GPUs tests we carried out is in Fig. 7.5. In the tests, we used four available high-end GPUs different as in chipset technology as in available onboard RAM:

- Quadro 5000, an industry-standard 40 nm Fermi series GPU of 2010 with 352 cores and 2.5 GB Graphics Double Data Rate 5 (GDDR5) [28];
- GeForce GTX 680, an entertainment 28 nm Kepler series GPU of 2012 with 1536 cores and 4 GB GDDR5 [29];
- GeForce GTX 970, an entertainment 28 nm Maxwell series GPU of 2014 with 1664 cores and 4 GB GDDR5 [30];
- TitanXp a top-end entertainment 16 nm Pascal series GPU of 2017 with 3840 cores and 12 GB GDDR5 [31].

With each GPU, we computed the FFT of a big single block data set produced by the Ultraview AD8-1500 $\times$ 2 DAQ board, so in the tests, we were limited only by GPU available onboard memory. With the algorithm we used, we realized that the computation was always completed in less than 350 ms having spare time both for other computations and for data movement to and from the accelerator.

From the chart in Fig. 7.3 we can see that the solution of performing the FFT on a single block data buffer posed a limit to the maximum signal bandwidth. However, it has guaranteed us to write more straightforward and faster software because we don't need to split the input bandwidth into sub-bands to analyze separately. Spectrometer state-of-art bandwidth split is carried out by mean of a Polyphase Filter Bank (PFB) because not produces a flat response across the channel, but also provides excellent suppression of out-of-band signals [32]. PFB is a typical computation-intensive DSP algorithm on all processors architectures, but that is high optimizable on FPGA. Not being able to have an accelerator with FPGA on our spectrometers because of the limited budget, the single block FFT analysis solution remained the most efficient available.

In general, considering that the processing carried out by digital spectrometers is organized in a pipeline in which, each block of acquired data, is subjected to different computations, the overall spectrometer performances are limited by the capabilities of the less performing module of the data processing chain. So, assembling a PC based spectrometer requires choosing COTS components between the ones available on the market to find a balanced set, without bottlenecks caused by the less performing part.

Moreover, in the COTS components research and selection, particular attention must be paid not only to compliance with the budget but also to assess the degree of standardization of the component. A careful assessment must be also be made on the possibility of recovering spare parts after years and on the availability in the future of more performing components, which, however, could be compatible with the current system.

For example, we chose to use Nvidia GPUs as computation accelerators, taking into consideration reliability, availability on the market over time, the functionality of the SDK, and its support as well. This choice has proved to be optimal over time for many aspects. The Nvidia GPU architecture improved about every two years (remaining anyway compatible with old PC motherboards). The SDK efficiently supports the GPUs produced in the last eight years without the need for substantial code rewrites but at least with only minimum fixes. Such stability was not present in other manufacturers' products, which in one case (Intel) withdrew the product from the market and has now replaced it with an FPGA-based accelerator.

7.2.4 Spectrometer's Software Design

In designing the spectrometer's software, we followed the same guidelines as in the hardware design, taking into consideration costs and performances. To limit costs, we decided to use open-source and free software together with ad-hoc developed application software. The overall spectrometer software is then composed of three main parts: the supporting OS, the processing application, and the control application.

Considering the NewSpec experience, the choice of the supporting OS was made taking into account:

- the COTS component driver availability and support,
- the software updates distribution and their impacts on the system,
- the support over time.

Following our previous positive experiences, it was chosen the Ubuntu Linux OS Long Term Support (LTS) distribution more compatible with the SDK (Software Development Kit) of both the GPU and the DAQ board.

The processing application is a Linux console application implemented in standard C/C++ with code regarding the GPU written with the Nvidia Compute Unified Device Architecture (CUDA) extensions. The application architecture is multi-threaded with every single task (data acquisition, data processing, data storage, system monitoring, etc.) managed by one or more different threads. Each task has a modular structure and can be configured in detail at the application startup, reading the configuration parameters on multiple configuration files. The application can be started by a console command, a shell script, or a scheduled operation, and its status can be monitored reading the operations and error log files.

The data processing is carried out following an operation chain that can be configured at the application startup. Each algorithm of the chain can be processed by the CPU or by the GPUs (if necessary and/or available) depending on the user selection. The processing results can be encoded in different data formats and saved on the spectrometer mass memory or sent out by a Network Interface Card (NIC).

An example of a typical SETI processing chain with GPU usage is the following:

1. data acquisition;
2. transfer of the acquired data to RAM;
3. synchronization with the GPUs and transmission of the data to be processed;
4. FFT execution (or other spectrum analysis algorithms if available);
5. boxcar (or other algorithms) execution to reject the filter shape from the spectrum and other operations on the data;
6. threshold triggering control (with average, moving average or other algorithms);
7. computation result formatting based on the requested output format.

The control application is written in Python to take advantage of language GUI portability and standardization, as well as its ability to interface with standard C/C++ applications easily.

This application must be used to configure, activate, stop, and control the processing application. The GUI is organized in different windows, each one configuring or controlling a different aspect of the processing application. The resulting setup is written on different files red by the processing application at its startup. Moreover, the control application monitor, the processing application status reading its output file, and the other information sent back.

Examples of available functionality in the control application are:

- auto-optimization of the processing application based on hardware components;
- dynamic construction of the processing chain;
- configuration of data acquisition and processing parameters;

- configuration of the station facilities in use;
- selection of the output data formats;
- monitoring and diagnostics of the system status;
- offline visualization of data output files.

Both the processing and the control application were developed to be easily adapted to other or future hardware components as well to be expanded with new functionality for other researches or applications.

7.3 The New Spectrometer Line

In the starting phase of the new spectrometer line, the NewSpec hardware was not available. Even if the system was out of order as a spectrometer, it was working as a high-end PC, and its computing power was used in other researches with very tight deadlines.

So it was decided to start building a new system with available spare or used components. This allowed the development of the pilot projects to carry out the first software tests to choose the components to use for the final system.

7.3.1 First Pilot Project

The first pilot project was carried out in 2015 using old hardware already present in the laboratory (see Fig. 7.6). The supporting system was a used PC equipped with an Intel i5 650 dual-core CPU, 4 GB RAM, and an Nvidia Quadro 5000 GPU [28]. The data acquisition was performed by a Picoscope 3204 digital oscilloscope connected by a High-Speed USB interface working up to 50 MS/s at a 12-bit resolution [34].

Tests carried out on the Picoscope showed an excellent RFI rejection, but its internal memory buffer of 256 kS limited the spectrometer performances at a maximum of 128 k channels (that we rounded up to 125 k). The memory buffer dimension allowed only a minimum frequency resolution of 200 Hz at the maximum signal bandwidth of 25 MHz with a spectra rate of 24 spectra per second. Moreover, this instrument cannot manage a continuous data flux because of the delay between successive acquisition introduced by its software drivers. Finally, further tests showed that the CPU efficiently processed a so reduced input data rate without delaying the other software operations. So we removed the GPU and converted the software to use the FFTW library returning to the NewSpec project system architecture (see Fig. 7.3).

This spectrometer was initially used as a test instrument dedicated both to software development, and Nvidia and AMD GPU tests. At the moment, it always connected to a cylindric parabolic antenna of the N/S arm of the Northern Cross Radio Telescope, and it is mainly used as a test system for software development and debug.

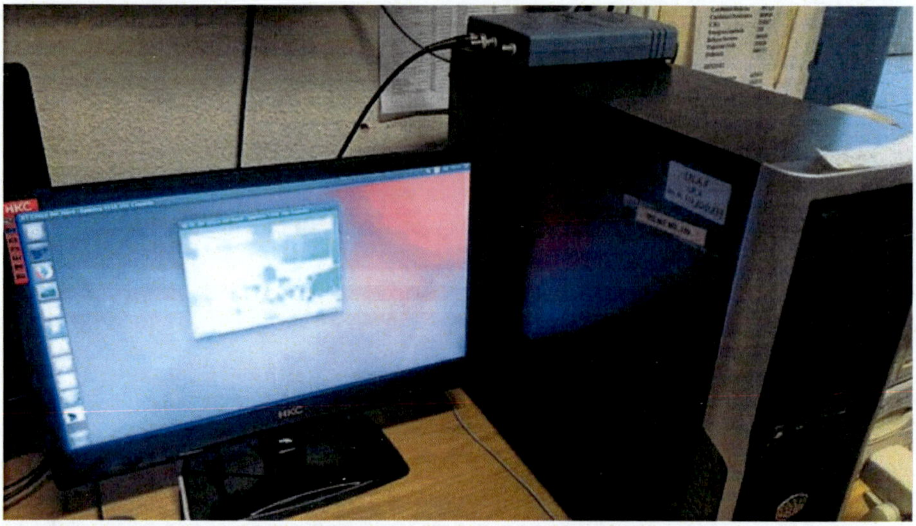

Fig. 7.6 Picture of the first pilot project spectrometer installed at the medicina radio astronomical station. The Picoscope DAQ is the light blue box on the PC case. On display, a window of the control application

7.3.2 Second Pilot Project

In the autumn of 2016, a second pilot project was carried out to develop a spectrometer with real-time analysis capabilities at 1 Hz resolution (see Fig. 7.7). Like the first pilot project, also this system was composed of obsolete COTS components that, in part, we recovered from eBay.

The system core was the DAQ board, an ADLink 9812A board. This board is a PCI adapter that can digitize four different signals at 20 MS/s with a 12-bit resolution [35]. This board can acquire data without any delay between two successive blocks and allow to manage signals with a maximum bandwidth of 10 MHz.

Tests of the boards, with the inputs shorted to the characteristic impedance, showed low resilience to RFI with the presence on the spectrum of sharp narrow lines at specific frequencies compatible with PC known clock signals frequency or their first harmonics (see Fig. 7.8).

System tests have proven that it meets the project specifications analyzing the entire bandwidth of 10 MHz in real-time at 1 Hz resolution. Still, but the DAQ board software drivers highlighted different incompatibility issues: the drivers don't support multi-core or Hyper-Threading (HT) CPUs, an installed RAM over 2 GB, and any Nvidia GPUs. We were forced to build the system around one of the last single-core Intel processors, the Pentium 4 516 at 2.93 GHz with a maximum power requirement of 84 W (so thermally stable) that had deficient computing capability than the one requested. It was able to output a 10 million channels spectrum with a delay of 300 s between each other.

Fig. 7.7 Picture of second pilot project internals. The ADLink DAQ board is the one on the center while the graphics adapter is on the left

Such a system was the second spectrometer with the same NewSpec project system architecture (see Fig. 7.3). It was initially used as a test bench for the application software development, and, despite the problems, it validated our technological strategies, letting us going ahead in the development of the final projects.

At the moment, this system is in stand-by waiting for the development of some ideas we had to raise its computing power.

7.3.3 The New Final Spectrometers

The DAQ boards turned out to be a critical issue in both pilot projects. Therefore, we decided to develop two new spectrometers in which we replaced these boards with new state-of-the-art devices.

The new DAQ boards allow analyzing signals with a much wider band without delay between successive spectra, but this required far more massive calculations than the single CPU design. Therefore, this type of spectrometer was designed according to the new hardware design described in Sect. 2.3 but with a small improvement. In the final instruments, we used two accelerator GPUs (see Fig. 7.9). In this design,

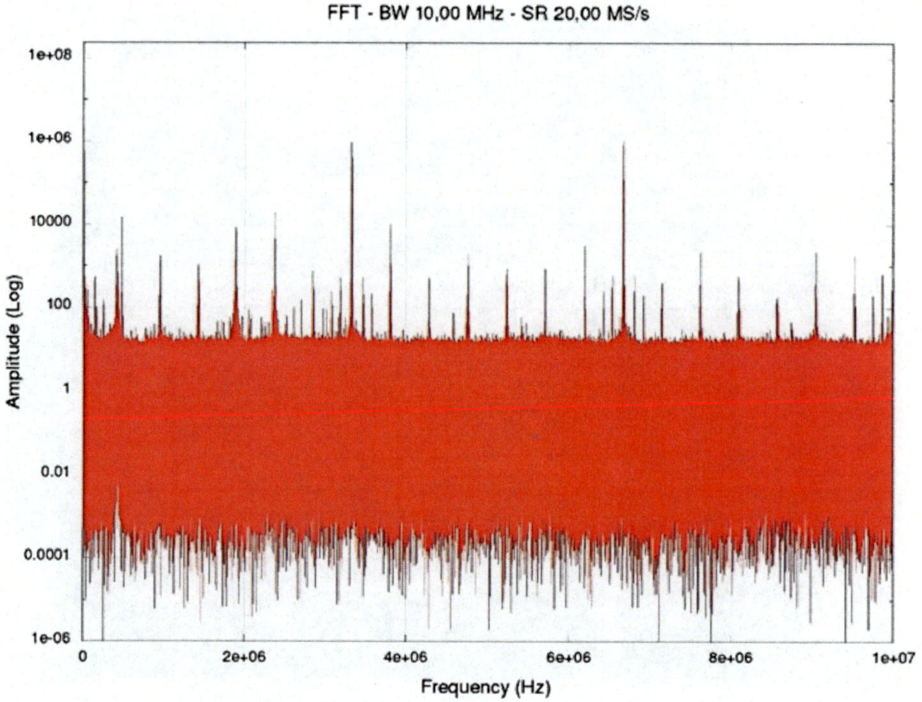

Fig. 7.8 Test spectra of the ADLink 9812A with input closed to characteristic impedance. In evidence, the strong narrow lines of RFIs

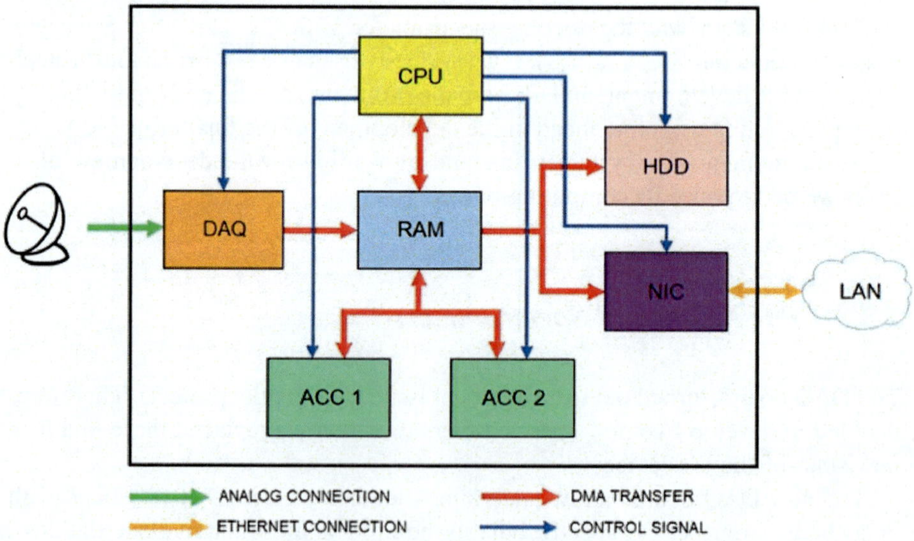

Fig. 7.9 Block diagram of the final spectrometer hardware design

all the data computations are charged to the GPUs, while the CPU has coordination tasks only.

The use of two GPUs was forced by the data transfer timings analysis that was of the same order of the FFT computation timings so very high time-consuming operation.

In this design, each data block produced by the DAQ board is transferred on a GPU that carry-out all the complete operation chains in a maximum time of 2 s. In the meanwhile, the next data block is transferred to the other GPU in an interleaved way. The GPU memory must contain all the data structures necessary for a single data block computation, accepting a maximum signal bandwidth proportional to the memory available. As shown in Fig. 7.4, a GPU with 4 GB of memory can analyze signals with a 150 MHz bandwidth, having 12 GB of memory let it possible to reach a 450 MHz bandwidth.

7.3.4 The First Final Project

In 2018, based on the first pilot project, we built a new spectrometer (see Fig. 7.10) that holds a USB oscilloscope, namely a Picoscope 3205D [36] with an 8-bit resolution, a 100 MHz maximum bandwidth, an internal memory for 250 MS and a USB 3.0 interface. In this case, like in the first pilot project, the real-time performances of this spectrometer are limited by the USB 3.0 bus, which, under optimal conditions, can exchange data at a maximum of 125 MS/s. Thus, the actual bandwidth of the

Fig. 7.10 Picture of the internals of the first final spectrometer. Inside the PC case, on the left, the GTX 680 GPUs, in the center the memory and the liquid-cooled CPU, outside the new Picoscope DAQ

spectrometer is limited to 50–60 MHz in a continuous acquisition mode, and to 100 MHz for non-continuous acquisition mode. The host system is a PC in stock from a past project composed by a liquid-cooled Intel quad-core i7-3770 K CPU, 16 GB Double Data Rate 3 (DDR3) RAM, and two high-end GPU NVIDIA GeForce GTX 680 with 4 GB of GDDR5 each [29].

Intensive tests have shown that this device can reach real-time spectra frequency resolution of 1 Hz for a 50 MHz bandwidth single-channel signals. In non-real-time applications, the system is only limited by the Picoscope bandwidth reaching spectra frequency resolutions of 1 Hz over a 100 MHz signal bandwidth.

Evaluating the possible application fields of this instrument, we realized that the input bandwidth limit of this spectrometer is too narrow for the requirements of actual INAF radio telescopes. With a total component cost of €3,000, this spectrometer can be useful for narrow bandwidth applications like NEO tracking or as a spectrometer for reduced bandwidth radiotelescopes like the ones used in amateur or educational applications.

7.3.5 The Second Final Project

In 2018, after careful market researches, we were able to identify a DAQ board suitable to equip the final system. The Spectrum M4i.2210- ×8 PCI Express DAQ board [37] can acquire signals data with a maximum sample rate of 1.25 GS/s on a single channel at 8-bit resolution (about 48 dB of dynamic range) with maximum signal bandwidth up to 500 MHz. Its software drivers let it works in continuous streaming with optimized data transfer to avoid bottlenecks in data transfer due to inefficient buses. Indeed, data transfers are performed in the background (DMA) without overloading the CPU and the PCI Express bus.

Starting from this DAQ board, and with the support of the NewSpec hardware, which in the meanwhile returned available, we built another spectrometer following the second pilot project experience (Fig. 7.11).

The components used for this spectrometer are the host system composed by a liquid-cooled Intel quad-core i7-3820 CPU, a 16 GB DDR3 RAM, a high-end motherboard that supports PCI Express bus with 40 lines, and by two high-end GPU Nvidia Titan Xp with 12 GB of GDDR5 each [31]. As we illustrated above, the size of the GPU memory is a crucial parameter because the dimension of a single block FFT analysis with CuFFT software library [23] depends on memory availability. From the tests we carried out with our software, 12 GB of GPU memory allowed us to perform FFT with 1 Hz resolution of signals with a bandwidth over 450 MHz.

Lab tests have shown that this spectrometer can reach a resolution of 1 Hz on a bandwidth of more than 450 MHz in real-time. The overall cost of this instrument is about €10,000, however, most of it (around 60%) being taken up by the DAQ board.

Fig. 7.11 Picture of the second new spectrometer during DAQ board tests. On the top of the PC case, there is an Ethernet network switch and a signal generator set to produce a 1 MHz simple sinusoidal signal (tone). On the PC display, the time domain signal with the spectrum in Fig. 7.12

7.4 Spectrometers Tests

7.4.1 Lab Tests

During the development phase, all spectrometers were tested in many ways. Taking apart the RFI resilience test that we described above, to check the correct behaviour of the spectrometers, we stimulated them with artificial signals. These signals vary from pure sinusoidal waves at a given frequency to more complex signals, including amplitude, frequency, and phase modulations. Usually, these signals are sent to the spectrometers over a bed of white electrical noise produced by a wideband noise generator. In our case, it was also simple to test our spectrometers with modified radiotelescope real signals.

At the Medicina Radio Astronomical Station, the spectrometers were installed in the receiver room in where is available the signal from a single cylinder (with four receivers just compensated in amplitude and phase to produce a single beam) of the N/S arm of the Northern Cross Radio Telescope.

This signal has a bandwidth of about 6 MHz with the typical profile of a flattened band-pass filter. Through a mixer, it is superimposed to the radio telescope signal,

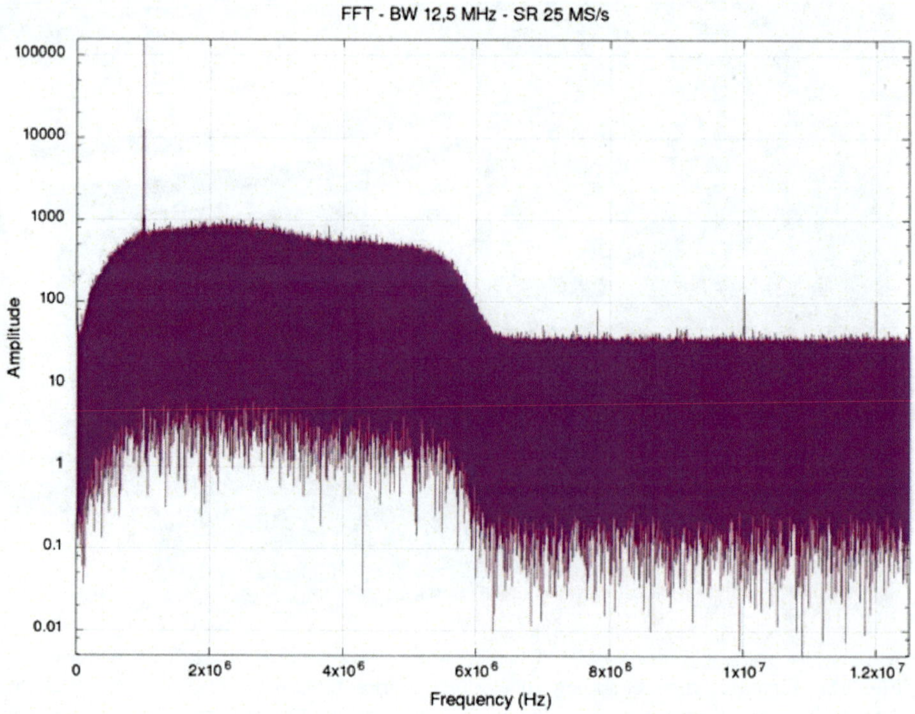

Fig. 7.12 The spectrum of a test signal shown in Fig. 7.11 composed by the Northern Cross Radio Telescope output signal with superimposed a 1 MHz tone

it is superimposed a sinusoidal signal at a given frequency, amplitude, and modulation (see Figs. 7.11 and 7.12). We used this kind of signals to make tests on the spectrometer sensibility and accuracy.

Moreover, signals allow us to detect RFI due to both spectrometer internals digital noise and local radio interferences.

Finally, we also used this kind of signal to test the spectrometers before operating the actual tests.

7.4.2 SETI Observation Tests

The first spectrometers operating test was carried out on September 20, 2018, when we officially obtained a 10 h slot (from 7:00 to 17:00 UT) to use the Medicina RT-32 for SETI observations on some habitable-zone exoplanet.

Our scheduled observation list included seven exoplanets that would be visible over the antenna horizon at that time slot. In the meanwhile, because of sudden antenna problems at the beginning of the time slot, our observation time was reduced to about a half, forcing the exclusion of various sources and the observation time

Table 7.1 September 20, 2018, SETI observation schedule and report of completed operations

Exoplanet	Number of observations	Observations length
Gliese 163 c	2	30′ and 50′
K2-18 b	1	2 h 10′
Kepler-174 d	None	
Luyten b	None	
Ross 128 b	1	35′
Trappist-1 d	None	
Wolf 1061 c	1	50′

reduction of others (see Table 7.1). Observations were carried out with a radio bandwidth centered at 1.42 GHz frequency (21 cm wavelength) named Hydrogen line.

Another problem related to these observations was the maintenance works to the Medicina RT-32 receivers room that was carried out in that month and reduced out work bandwidth to 20 MHz. So, based on all the problems faced, we decided to reduce the bandwidth to 10 MHz and to use both the first and the second pilot project spectrometers connecting them to the same signal with the secondary goal of comparing their results in the postprocessing phase [38, 39].

The spectrometer data processing was set to the standard Serendip IV processing (see Fig. 7.13). It is composed by a sequence of operations: FFT (a), normalization (d) composed by boxcar computation, and subtraction (b) to remove the filter shape followed by moving average (c) to compute the threshold, as spectrum local mean value multiplicated by a programmable factor), and final selection of the spectrum rows with amplitude over the threshold. The first pilot project spectrometer was set up at 12.5 MHz bandwidth with a 100 Hz resolution obtaining a spectra rate of about 10 spectra per second. The second pilot project spectrometer was set up at 10 MHz bandwidth with a 1 Hz resolution obtaining a spectrum every 5 min.

Observation results were saved as in two file format:

- SIV—the standard output file format of Sentinel IV spectrometer suitable for the use of the old Sentinel IV post-processing software,
- RAW—the standard new spectrometers line file format for general SETI observations (a more straightforward and improved version of SIV) (see Fig. 7.14).

The comparison of the observation results, produced by both spectrometers, proved a precise agreement between them (net of the technical parameters of the two spectrometers such as spectra resolution and rate).

Post-processing of observations results showed no results in ETI signal detection [40].

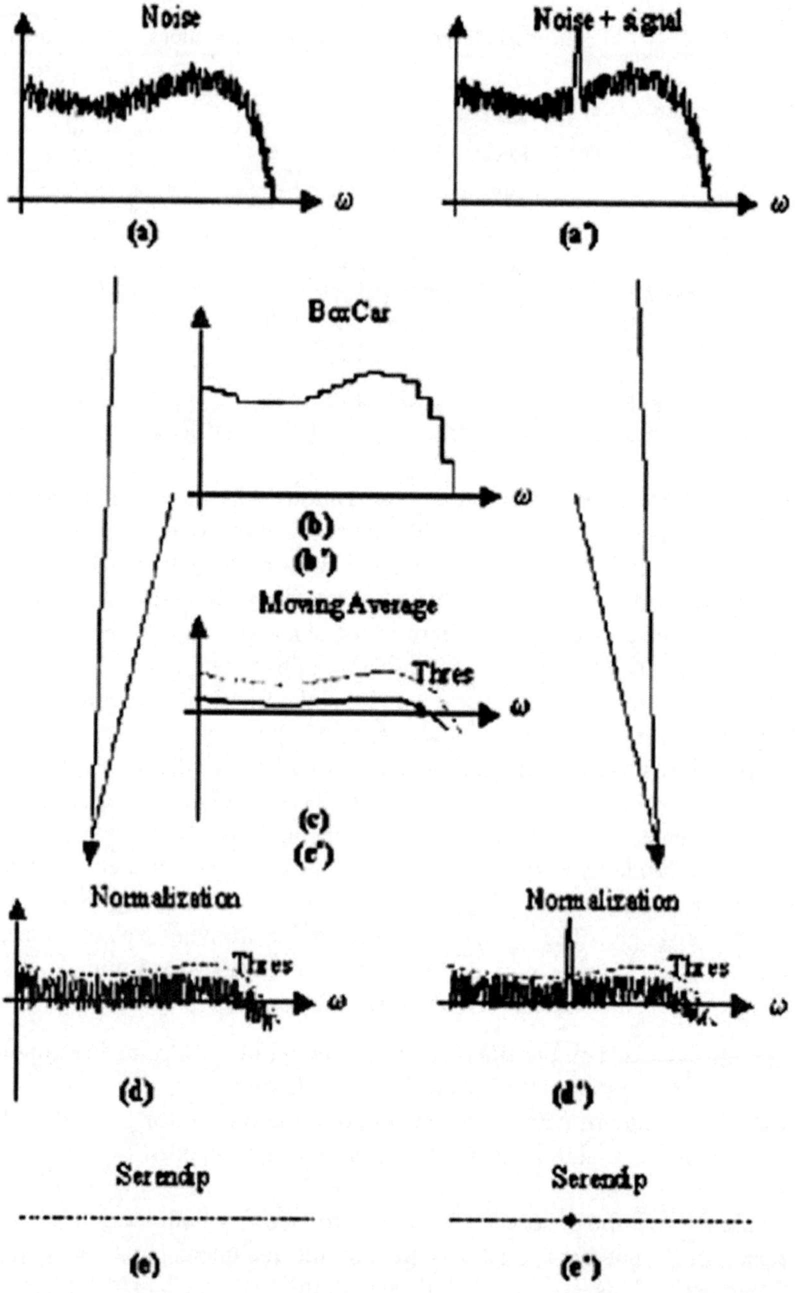

Fig. 7.13 Standard Serendip IV SETI processing chain (Credits to Stelio Montebugnoli, IRA—Medicina Radio Astronomical Station)

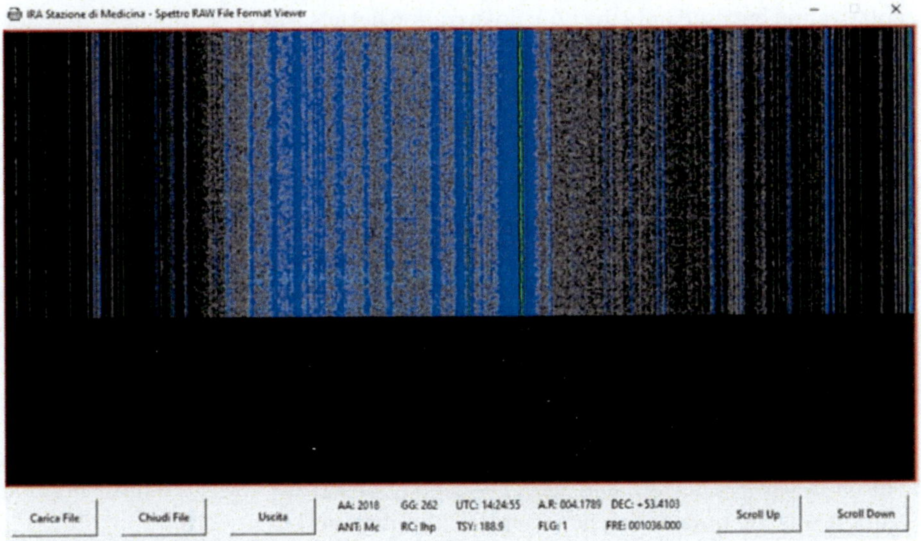

Fig. 7.14 Screenshot of the waterfall diagram produced by the RAW file format viewer/analyzer function of the control application. The data on the screen were read from a file produced during the SETI observation test

7.4.3 NEO Observation Tests

The second spectrometers operating test was the radar observation of the asteroid 2003 SD220 performed in December 2018.

The asteroid 2003 SD220 was discovered by Lowell Observatory Near-Earth-Object Search (LONEOS) on September 29, 2003, and the Minor Planet Center has classified it as a "Potentially Hazardous Asteroid" (PHA). 2003 SD220 has an extremely slow rotation period of roughly 12 days, and the lightcurves suggest non-principal axis rotation. It approached within 0.019 au on December 22, 2018, making it one of the strongest radar targets of the last years. For this reason, an international radar campaign was planned to observe this asteroid in the days around its close approach to Earth using a bistatic radar configuration. Some Continuous Wave (CW) unmodulated transmission runs, were performed in X-band (8560 MHz) by the Goldstone DSS-14 antenna. In this configuration, some radio telescopes, such as the Italian radio telescopes SRT (64-m dish) and Medicina RT-32, were used as receiver part of a bistatic radar system. Due to the slow rotation rate of the 2003 SD220, the received echoes from the asteroid were quasi-monochromatic signals, so they were an ideal test bench for the new spectrometer line instruments.

Different time intervals were selected for the asteroid observation taking into account both the common visibility window of the asteroid 2003 SD220 from Medicina RT-32 and Goldstone DSS-14 as well the limits in pointing angle of both antennas. The Goldstone DSS-14, a 70-m fully steerable dish, radiated a CW signal in RCP polarization. In contrast, the Medicina RT-32 received both LCP and RCP polarization but used only the LCP polarization because the spectrometers were set to work on a single channel. In this polarization, opposite to that transmitted, we had the maximum power of the signal reflected by the asteroid surface. During our observations, the frequency of the transmitted signal was continuously tuned to compensate for the expected Doppler shift between the Goldstone DSS-14 antenna and the SRT 64-m radio telescope.

For these observations, we used the first pilot project and the first final project for the echo radar detection. We also used the second pilot project but only for comparative tests because of its too low spectra rate. The signal bandwidth received from the Medicina RT-32 was filtered to 781.250 kHz and the spectrometers were set up at a frequency resolution of 6.25 Hz useful for maintaining the residual Doppler signal smearing inside a single frequency channel.

A preliminary test was performed on December 11 by injecting a signal similar to the expected radar echo into the spectrometer. The NEO observations were carried out on December 17, 20, and 22. The last day observation was the most important since it was the time of the close approach of the asteroid to Earth.

The first two observations were made controlling the spectrometers from the receiver room, while the last observation was made controlling the spectrometers by remote. The spectrometer data were stored in binary files with a dedicated format (named NEO) and analyzed by post-processing software developed specifically in MATLAB. Although the spectra appear to be contaminated by a high noise level, which is still under investigation, the radar echoes from the asteroid were successfully detected at the expected frequency. Figure 7.15 shows the observed radar echo, integrated for about 10 s, acquired on December 22 at 14:47:23 UT.

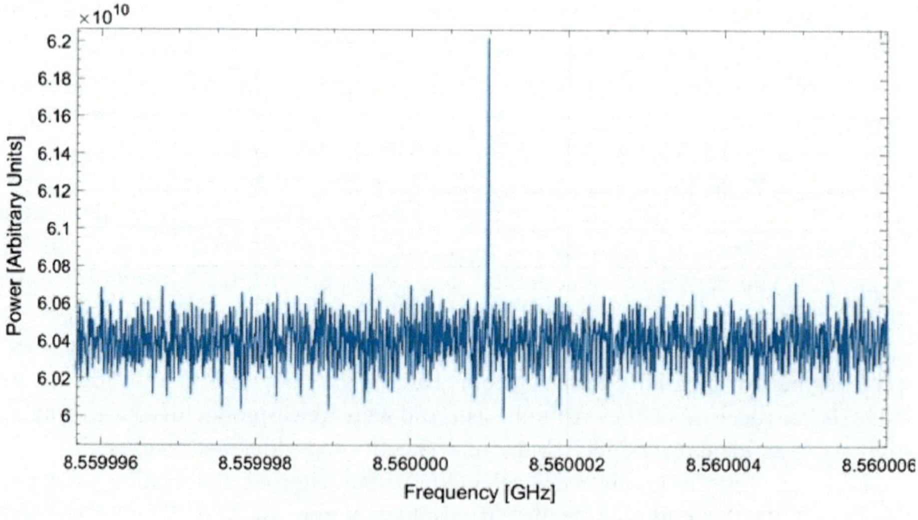

Fig. 7.15 The asteroid 2003 SD220 radar echo detected in LCP polarization on December 22nd at 14:47:23 UT. It was produced integrating 54 single spectra obtained at 6.25 Hz frequency resolution

7.5 Conclusions

The tests described above showed very promising results that led us to continue the development of the spectrometer line.

Both pilot projects, that have proven to be very useful in the design definition phase, and the initial software development, will continue to be used for the same purposes but also in applications other than SETI. The first final project will be optimized for narrowband scientific research (NEO, space debris, and so on). Its development seems to be very interesting for amateur and educational applications Finally, the second final project will be optimized strictly for broadband (radio telescope IF) SETI researches in piggyback operations.

Like in past experiences, SETI technical researches and related instruments proved to be very useful also in other research contexts.

7.6 Possible Future Developments

Many future developments of the new spectrometer line both in the hardware processing capabilities and in the application software algorithms and functions are possible.

Regarding the first pilot project, we will carry out comparative evaluations on different algorithms like KLT, HHT, Cognitive Radio, Agnostic Entropy, etc. under the same input signal, both artificially generated and detected. Maybe we shall try KLT optimizations on other computation accelerators.

For what concerns the second pilot project, the priority is raising its computing performances. To get this result, we will follow two approaches. In the first approach, we will try to use AMD GPUs with OpenCL to acquire experiences on an SDK that is useful in the future if the prices of FPGA accelerators will drop. In the second approach, we will set up a low-cost computer cluster (see Fig. 7.16) composed of ARM-based boards (Raspberry PI, UDOO QUAD, Pine64, Nvidia Jetson, etc.). ARM architectures, in the last years, showed promising results in HPC applications [41] also because they are implemented with a better lithographic process than Intel CPUs.

The first final project will probably be used in more specialized NEO observations, optimizing their functionality for this purpose. Other exciting applications of this spectrometer will be dedicated software versions for amateur and educational fields.

The second final project will be upgraded with newer GPUs (Nvidia Volta, Turing, Volta, or Ampere family) to let users to use the entire DAQ available bandwidth and to allow more sophisticated computations.

Fig. 7.16 Preliminary rendering of a 24 Raspberry Pi compute module [42] cluster board (Credits to Enrico Svampa, Kunst Engineering S.r.l.)

Disclaimer All trademarks present belong to their legitimate owners; third party brands, product names, trade names, corporate names, and companies mentioned may be trademarks of their respective owners or registered trademarks of other companies and have been used for the purpose of explanation and for the benefit of the owner, without any violation of the copyright rights in force.

References

1. S. Montebugnoli, C. Cosmovici, G. Bianchi, G. Naldi, J. Monari, S. Pluchino, M. Bartolini, E. Salerno, G. Pupillo, F. Schilliró, A. Mattana, SETI back end made inexpensive, in *International Astronautical Federation (IAF) (eds.), Proceedings of 61st International Astronautical Congress 2010 (IAC 2010), IAC-10.A4.1.4*, vol. 5 (2010), pp. 4284–4288
2. S. Montebugnoli, M. Cecchi, J. Monari, E. Spisni, Sviluppo di software per uno spettrometro digitale di segnali radio basato su scheda DSP C62. Note IRA 284 (1999)
3. C. Bortolotti, M. Cecchi, S. Montebugnoli, M. Roma, Sentinel 2: sistema per il monitoraggio delle radio-interferenze. Note IRA 294 (2000)
4. S. Mariotti, S. Montebugnoli, J. Monari, C.-T. Langvik, P. Norli, EMBLA 2000 Project Plan. Note IRA 283 (2000)
5. G. Moore, Cramming more components onto integrated circuits. Electronics **38**(8), 114 (1965)
6. G. Amdahl, Validity of the single-processor approach to achieving large scale computing capabilities, in *Proceedings of the April 18–20, 1967, Spring Joint Computer Conference, Series AFIPS '97 (Spring)* (1967), pp. 483–485. https://doi.org/10.1145/1465482.1465560
7. A.-S. Tanenbaum, H.-J. Bos, *Modern Operating Systems*, 4th edn. (Pearson Higher Education, Chap. 8, 2015)
8. NVIDIA homepage, https://www.nvidia.com. Accessed on 24 June 2020
9. Wikipedia Xeon Phi web page, https://en.wikipedia.org/wiki/Xeon_Phi. Accessed on 24 June 2020
10. NVIDIA developer homepage, https://developer.nvidia.com. Accessed on 24 June 2020
11. Intel Math Kernel Library web page, https://software.intel.com/content/www/us/en/develop/tools/math-kernel-library.html. Accessed on 24 June 2020

12. Wikipedia FPGA web page, https://en.wikipedia.org/wiki/Field-programmable_gate_array. Accessed on 24 June 2020
13. Intel FPGA OpenCL web page, https://www.intel.it/content/www/it/it/software/programmable/sdk-for-opencl/overview.html. Accessed on 24 June 2020
14. Xilinx FPGA OpenCL web page, https://www.xilinx.com/html_docs/xilinx2017_4/sdaccel_doc/lyx1504034296578.html. Accessed on 24 June 2020
15. S. Montebugnoli, C.-B. Cosmovici, J. Monari, SETI-Italia, in *Bioastronomy 99, Astronomical Society of the Pacific Conference Series (ASPC)*, vol. 213 (2000), pp. 501
16. M. Frigo, S.-G. Johnson, The Fastest Fourier Transform in the West. Technical Report MIT-LCS-TR-728 (1997)
17. M. Frigo, S.-G. Johnson, FFTW: an adaptive software architecture for the FFT, in *Proceedings of the 1998 IEEE International Conference on Acoustics, Speech and Signal Processing, ICASSP'98*, vol. 3 (1998), pp. 1381–1384. https://doi.org/10.1109/ICASSP.1998.681704
18. M. Frigo, S.-G. Johnson, The design and implementation of FFTW3. Proc. IEEE **93**(2), 216–231 (2005). https://doi.org/10.1109/JPROC.2004.840301
19. Ultraview AD8-1500x2 user manual, https://www.ultraviewcorp.com/manuals/PCIe_DAQ_Manual_071113-0.pdf. Accessed on 24 June 2020
20. R. Lulli, G. Bianchi, J. Monari, S. Montebugnoli, Low cost SETI data multi-processing, in *International Astronautical Federation (IAF) (eds.), Proceedings of the 69th International Astronautical Congress (IAC 2018) IAC-18.A4.1.5* (2018), pp. 2377–2382
21. ROACH board web page, https://casper.berkeley.edu/wiki/ROACH. Accessed on 24 June 2020
22. ROACH2 board web page, https://casper.berkeley.edu/wiki/ROACH2. Accessed on 24 June 2020
23. NVIDIA CuFFT toolkit documentation, https://docs.nvidia.com/cuda/cufft/index.html. Accessed on 24 June 2020
24. J. Wu, J. JáJá, Optimized FFT computations on heterogeneous platforms with application to the Poisson equation. Journal of Parallel and Distributed. Computing **74**(8), 2745–2756 (2014). https://doi.org/10.1016/j.jpdc.2014.03.009
25. G. Teodoro, T. Kurc, J. Kong, L. Cooper, J. Saltz, Comparative performance analysis of Intel Xeon Phi, GPU, and CPU: a case study from microscopy image analysis. IEEE Trans. Parallel Distrib. Syst. 1063–1072 (2014). https://doi.org/10.1109/IPDPS.2014.111
26. J. Fang, A. Varbanescu, B. Imbernón, J. Cecilia, H. Pérez-Sánchez, Parallel computation of non-bonded interactions in drug discovery: Nvidia GPUs vs. Intel Xeon Phi, in *Proceedings of the 2nd International Work-Conference on Bioinformatics and Biomedical Engineering (IWBBIO'14)* (2014)
27. T. Liu, S.-G. Xu, C.-D. Carothers, Comparison of two accelerators for Monte Carlo radiation transport calculations, Nvidia Tesla M2090 GPU and Intel Xeon Phi 5110p coprocessor: a case study for X-ray CT imaging dose calculation. Ann. Nucl. Energy **82**, 230–239 (2015). https://doi.org/10.1016/j.anucene.2014.08.061
28. NVIDIA Quadro 5000 brochure, https://www.nvidia.com/docs/IO/40049/NV-DS-QUADRO-5000-Jul10-LR-final.pdf. Accessed on 24 June 2020
29. NVIDIA GeForce GTX680 web site, https://www.geforce.com/hardware/desktopgpus/geforce-gtx-680. Accessed on 24 June 2020
30. NVIDIA GeForce GTX970 web site, https://www.nvidia.com/it-it/geforce/900-series/. Accessed on 24 June 2020
31. NVIDIA Titan Xp web site, https://www.nvidia.com/en-us/titan/titan-xp/. Accessed on 24 June 2020
32. CASPER Polyphase Filter Bank web page, https://casper.ssl.berkeley.edu/wiki/The_Polyphase_Filter_Bank_Technique. Accessed on 24 June 2020
33. NVIDIA CUDA toolkit documentation homepage, https://docs.nvidia.com/cuda/cuda-c-programming-guide/index.html. Accessed on 24 June 2020
34. Picoscope 3204 brochure, https://www.picotech.com/download/datasheets/picoscope-3204-3205-3206-datasheet.pdf. Accessed on 24 June 2020

35. ADLink PCI-9812/9810 website, https://www.adlinktech.com/Products/GPIB_Modular_Instruments/Digitizers_Oscilloscopes/PCI-9812_9812A_9810?lang=en. Accessed on 24 June 2020
36. Picoscope 3205D web site, https://www.picotech.com/oscilloscope/3000/picoscope-3000-oscilloscope-specifications. Accessed on 24 June 2020
37. Spectrum M4I.2210-x8 web site, https://spectrum-instrumentation.com/de/m4i2210-x8, last accessed 2020/06/24.
38. S. Montebugnoli, J. Monari, F. Ravaglia, S.A.L.V.E. Software Aimed at off Line Verification Eti software for Serendip data—Ver. beta 0. Note IRA 280 (1999)
39. J. Monari, S. Montebugnoli, A. Cattani, M. Maccaferri, S. Mariotti, G. Maccaferri, A. Scalambra, M. Cecchi, A. Orlati, Spettrometro ad altissima risoluzione Serendip IV: algoritmi di mitigazione delle interferenze e software di post elaborazione. Note IRA 299 (2000)
40. SETI Institute ETI Signal Detection web page, https://www.seti.org/protocols-eti-signal-detection. Accessed on 24 June 2020
41. N. Rajovic, A. Rico, F. Mantovani, D. Ruiz, J. Oriol Vilarrubi, C. Gomez, L. Backes, et al., The Mont-Blanc prototype: an alternative approach for HPC systems, in *SC '16: Proceedings of the International Conference for High Performance Computing, Networking, Storage and Analysis* (2016), pp. 444–455. https://doi.org/10.1109/SC.2016.37
42. Raspberry Pi Compute Module datasheet, https://www.raspberrypi.org/documentation/hardware/computemodule/datasheets/rpi_DATA_CM3plus_1p0.pdf. Accessed on 24 June 2020

Chapter 8
Moon Farside Protection, Moon Village and PAC (Protected Antipode Circle)

Claudio Maccone

Abstract The Moon Farside is the only place in space, and not too far from the Earth, where radio transmissions and noises produced by Humanity on Earth may not reach since the spherical body of the Moon blocks them, acting like a shield. Thus, protecting the Moon Farside from all kinds of non-scientific future exploitations (e.g. real estate, industry and military) has long been a concern for many far-sighted space scientists as well as for several IAA Academicians. We started facing this problem in the 1990s, when the French radio astronomer Jean Heidmann of the Paris Meudon Observatory first promoted an IAA Cosmic Study about which areas of the Moon Farside should be reserved for scientific uses only. But Heidmann passed away on July 2, 2000, and his work had to be continued by others. This author took over his IAA Cosmic study and a paper describing both the scientific and legal aspects of the problem was published in 2008. Later, on June 10, 2010, this author was the first scientist to present the case for the Moon Farside Protection at the United Nations Office of Outer Space Affairs in Vienna during a meeting of UN-COPUOS, the United Nations Committee on the Peaceful Uses of Outer Space. Unfortunately, the undeclared but quite real "current, new race to the Moon" complicates matters terribly. All the space–faring nations now keep their eyes on the Moon, and only the United Nations might have a sufficient authority to Protect the Farside and keep safe its unique "radio-noise free" environment. But time is money, and the "Moon Settlers" may well reach the Moon before the United Nations come to agree about any official decision concerning the Farside Protection. Quite an URGENT ISSUE. In this paper, we propose that the new "Moon Village" supported by the vision of the ESA Director General, Jan Woerner, be located OUTSIDE the PAC (Protected Antipode Circle) obviously not to interfere with the detection of radiation coming from space, but also SOUTH OF THE PAC, to be "close" to the South Pole as much as needed to benefit of water there. It thus appears the best venue for the "Moon Village" would be on or around the 180 degree meridian and south to the −30 degree in latitude of the PAC, possibly much more south of that, almost at the South Pole, thus resolving Moon Village VENUE ISSUE.

C. Maccone (✉)
International Academy of Astronautics (IAA) and Istituto Nazionale di Astrofisica (INAF), Torino (Turin), Italy
e-mail: claudio.maccone@gmail.com

S. Montebugnoli and A. Melis (eds.), *The Search for ExtraTerrestrial Intelligence*, Springer Proceedings in Physics 260, https://doi.org/10.1007/978-3-030-63806-1_8

8.1 History

On June 10th, 2010, this author made a presentation in front of the United Nations Committee on the Peaceful Uses of Outer Space (COPUOS) in Vienna. His presentation was of course archived by the Office of Outer Space Affairs (OOSA) and is freely downloadable at the web site https://www.unoosa.org/pdf/pres/copuos2010/tech-06E.pdf.

He made this presentation on behalf of the International Academy of Astronautics (IAA, based in Paris) and the Fig. 8.1 shows him while making his presentation, while Fig. 8.2 shows his United Nations badge.

This presentation's goal was to ask the United Nations to *protect* a circular piece of land on the Moon Farside centered around the Antipode (the point opposite to the Earth at the Farside center). The word "protect" means here to reserve that "Protected Antipode Circle (PAC)" for use by scientists only, especially radio astronomers, forbidding any other non-scientific activity by realtors, industry, tourists and the military. A neat description of this project was recently re-published in Reference [1], and will not be repeated here. Previous work by this author and his co-workers Salvo Pluchino and Nicolò Antonietti are described by the References [2–5]. However, our readers are strongly advised to read those papers carefully, just to know the reasons why they, and the International Astronomical Union (IAU) in particular, should support this author's project before it gets too late. Thanks.

Fig. 8.1 Claudio Maccone while making his presentation at the IAA Seat in COPUOS, the United Nations Committee on the Peaceful Uses of Outer Space, on June 10, 2010 in Vienna, Austria. That was possible since the IAA is a Non-Governmental-Organization (NGO) recognized as such by the United Nations

Fig. 8.2 Claudio Maccone's badge enabling him to attend the COPUOS Meetings in Vienna, June 9–18, 2010, and make his presentation on June 10, 2010

One might naïveley expect that, after such an official presentation at the United Nations, something should have happened towards the Moon Farside's protection, either at the political level or the scientific level or at both. But you are to be disappointed: nothing happened at all for the next five years after 2010.

Why?

Because no national "big" space agency, nor any private entrepreneur had the technological capability of sending a spacecraft to the Moon Farside before until about 2015. The International Astronomical Union should have been the #1 promoter of the Farside Protection didn't care either, or was just unaware even of the urgency of this issue: they love too much topics like "dark matter", or what will happen when the Andromeda Galaxy collides with the Milky Way in the 3 billion years, and similar "learned stuff", to "get down" to legal-scientific topics like the Moon Farside Protection.

The situation started to change only in the summer of 2015, when the new Director General of ESA, Prof. Jan Woerner, began his term. Woerner declares himself to be a civil engineer, very much interested in building a Moon Village possibly on the Farside, though the Moon Village location was not specified yet.

Next, a host of eager "Moon bases constructors" established the Moon Village Association (November 2017).

But nobody cared to ask to reserve the central part of the Farside, specifically crater Daedalus, for usage by radio astronomers only. Again this author's voice was a voice in the desert, once again.

8.2 The Year 2018

The year 2018 has seen a couple of innovative and politically quite important Chinese space missions (for more details, please see the Wikipedia site https://en.wikipedia.org/wiki/Chang%27e_4#Queqiao_satellite, from which we now freely quote):

(1) Direct communications with Earth are impossible on the Farside of the Moon, since transmissions are blocked by the Moon itself. Communications must go through a communications relay satellite, which is placed at a location that has a clear view of both the Farside landing side and the Earth. On May 20, 2018, the Chinese National Space Agency (CNSA) launched the Queqiao ("Magpie Bridge") relay satellite to a halo orbit around the Earth–Moon L2 Lagrangian point. The relay satellite has a mass of 425 kg, and it uses a 4.2-m antenna to receive X band signals from the lander and rover of the Chang'e 4 space mission to the Farside and relay them to Earth control on the S band. This spacecraft took 24 days to reach L2, using a lunar swing-by to save fuel. On June 14, 2018, Queqiao finished its final adjustment burn and entered the L_2 halo mission orbit, which is about 65,000 km above the Moon Farside. This is the first lunar relay satellite ever at this location, but there might be more in the future, when other space-faring nations decide to explore the Farside.

(2) The Chang'e 4 spacecraft, including a robotic lander and a rover, was launched on December 7, 2018, and landed on the Farside on January 3, 2019, in the Von Kármán crater (180 km in diameter), that is in the South Pole-Aitken Basin on the Moon Farside. These two space missions are marking an historic change in the Moon exploration. In fact, not only this is the first time that the exploration of the Moon Farside is attempted, but the Moon Farside "radio silence" that existed prior the Queqiao arrival at L2 is now partially polluted for the first time by the radio emissions of Queqiao in both X and S bands.

(3) As we mentioned already, the International Academy of Astronautics (IAA, based in Paris) had been studying since the 1990s the possibility of setting up a future radio telescope inside crater Daedalus, an 80 km crater practically located at the center of the Farside. One might also say that Daedalus is very close to the Antipode, the point along the Earth-Moon axis exactly opposite to the Earth on the Farside. On 10 June 2010, this author presented at the United Nations Committee on the Peaceful Uses of Outer Space (COPUOS, based in Vienna) his proposal to create a "Protected Antipode Circle" (PAC) area around the Antipode to let a future radio telescope landed inside Daedalus to observe the radio sky as "purely" as it was before Humans discovered radio waves. In other words, this author's proposal to the UN was to ask the space-faring nations

to come to an agreement reserving the PAC for astronomical exploration only, since crater Daedalus is the closest location to Earth where the radio pollution produced by Humankind does not still practically reach. This activity would be especially important for SETI, the Search for Extra Terrestrial Intelligence that NASA now calls "Searching for Technosignatures" just to avoid the "too-political" word SETI. Actually, setting up more than one radio telescope inside Daedalus would enable interferometry, the best technique for radio astronomical searches ever. Even more precisely, these radio telescopes would not be dishes at all: they would be phased arrays, i.e. planar deployable surfaces with dipoles commanded by computers, of course much easier to be landed inside Daedalus than big dishes.

(4) However, the communications relay satellite Queqiao now at L2 changes this idyllic picture. In fact, it is obvious that the future astronomical observatory Daedalus will have to abandon radio exploration on the X and S bands because blinded by Queqiao, unless negotiations between China and the other space-faring nations will reach a general agreement on which bands are reserved for what. This is a task for international organizations like ITU (International Telecommunications Union), the American CORF (Committee on Radio Frequencies) and the European CRAF (Committee of Radio Astronomy Frequencies), plus other related organizations worldwide, like the Moon Village Association (MVA).

(5) On March 27, 2019 this author, on behalf of the IAA, intends to organize in Paris an IAA International Conference by the title of "Moon Farside 2019": the first attempt ever to reach an international agreement about the frequencies allowed or not allowed over the PAC in the future.

8.3 The Farside Spectrum Still is Not Polluted (in December 2020) Except in the S, X and UHF Bands

Let us now take a closer look at the electromagnetic radio spectrum as given, for instance at the Wikipedia site https://en.wikipedia.org/wiki/Radio_spectrum from which the two Figs. 8.3 and 8.4 are taken.

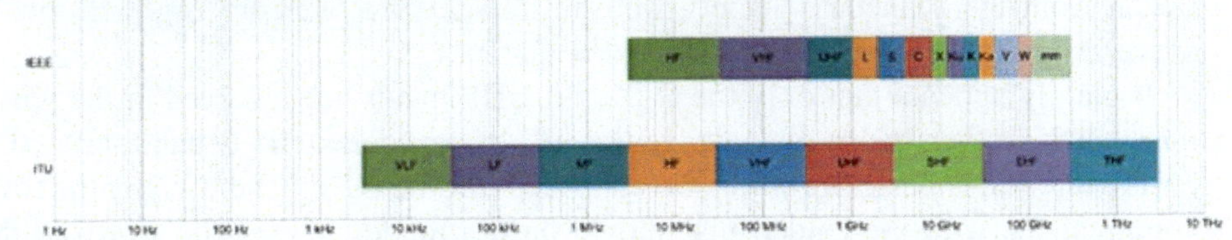

Fig. 8.3 Comparison of radio band designation standards

Radar-frequency bands according to IEEE standard[7]		
Band designation	**Frequency range**	**Explanation of meaning of letters**
HF	0.003 to 0.03 GHz	High Frequency[8]
VHF	0.03 to 0.3 GHz	Very High Frequency[8]
UHF	0.3 to 1 GHz	Ultra High Frequency[8]
L	1 to 2 GHz	Long wave
S	2 to 4 GHz	Short wave
C	4 to 8 GHz	Compromise between S and X
X	8 to 12 GHz	Used in WW II for fire control, X for cross (as in crosshair). Exotic.
K_u	12 to 18 GHz	Kurz-under
K	18 to 27 GHz	Kurz (German for "short")
K_a	27 to 40 GHz	Kurz-above
V	40 to 75 GHz	
W	75 to 110 GHz	W follows V in the alphabet
mm or G	110 to 300 GHz [note 1]	Millimeter

Fig. 8.4 Radio frequency bands according to IEEE standards

Figure 8.4 shows the **IEEE radar bands**. Frequency bands in the microwave range are designated by letters. This convention began around World War 2 with military designations for frequencies used in radar, which was the first application of microwaves. Unfortunately there are several incompatible naming systems for microwave bands, and even within a given system the exact frequency range designated by a letter varies somewhat between different application areas. One widely used standard is the IEEE radar bands established by the US Institute of Electrical and Electronic Engineers.

8.4 Queqiao and Chang'e 4 Communications Bands Above the Moon Farside

Please look at Fig. 8.5, summarizing well the telecommunications involved with the Chang'e 4 and Queqiao spacecrafts, as well as with the rover exploring the Farside (taken from the Planetary Society website https://www.planetary.org/multimedia/space-images/spacecraft/change-4-mission.html).

It plainly appears that only the S, X and UHF bands are involved in the whole mission. No other bands are involved at all. In particular, the L band, crucial to SETI searches, is basically *not* involved, except for the UHF band used by the rover communicate with the lander, that should have a modest power (unknown to this author at the moment). Thus, had we radio telescope nowadays working inside crater

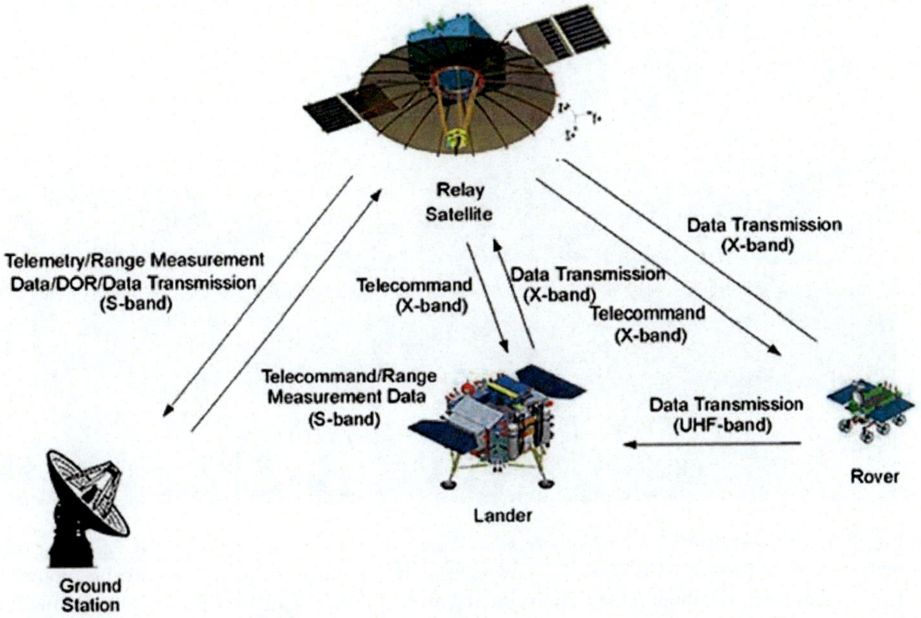

Fig. 8.5 Telecommunications involved with the Chang'e 4 and Queqiao spacecrafts

Daedalus for SETI, the Chinese Farside spacecrafts would NOT hamper its searches for ETs at all: good news for SETI scientists.

8.5 COSMOLOGY: Need for Ultra-Low Frequency Radio Astronomy in Space Within the Quiet Cone Above the Moon Farside, I.E. Just at the Lagrangian Point L2, Where Queqiao Is

Let us now consider ultra-low frequency radio astronomy (frequencies below ~20 MHz). At these frequencies, "single dish" (or "single phased array", if you wish) radio astronomy is quite helpless due to very low angular resolution: it will be totally knocked-out by the confusion effect. The only way out is to employ interferometry. However, employing radio *interferometry* on the Farside will bring its own issues in the technology. The case in question is well exemplified by Cosmology, as we now briefly describe.

NCLE (Netherlands-China Low Frequency Explorer) is considered a pathfinder mission for a future low-frequency space-based or moon-based radio interferometer which has the detection and tomography of the 21-cm Hydrogen line emission from the Dark Ages period as the principal science objective. In simple words, the hydrogen line that nowadays is at 1420 MHz, i.e. 21 cm, was at a much lower frequency about

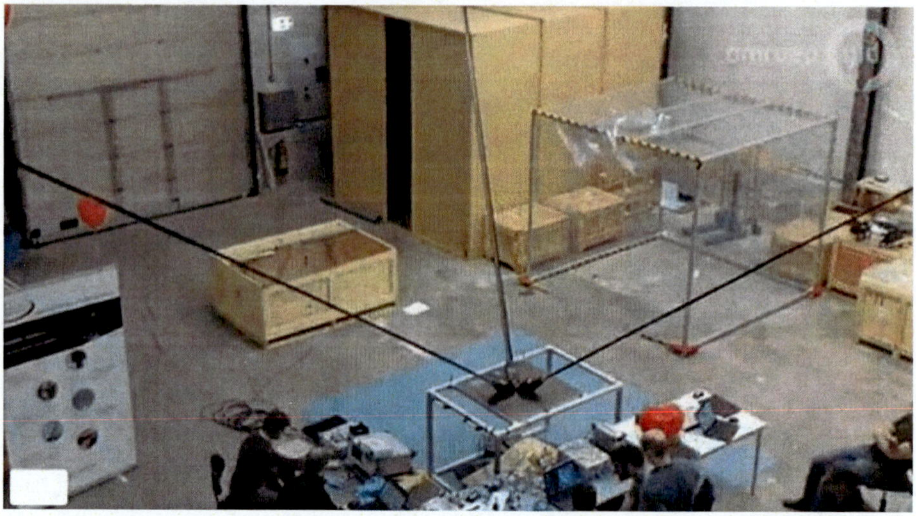

Fig. 8.6 **COSMOLOGY**: the Netherlands-China Low-frequency Explorer (NCLE) instrument with three 5-m antennas for Queqiao, now (December 2020) in place in a halo orbit around the Lagrangian point L2 about 65,000 km right above the Farside. This NCLE is essentially made by three 5-m-long ribs (very low frequency antennas) here shown under construction in the Netherlands before they were sent to China for the Queqiao launch, in the spring of 2018. Principal Investigator of NCLE is renowned astronomer Heino Falcke, sponsored by ESA to build NCLE in the Netherlands

300,000 years after the Big Bang, and cosmologists hope to detect it at L2 by virtue of NCLE (Fig. 8.6).

8.6 SETI (Or "Technosignatures", According to NASA's 2018 "New Jargon")

SETI, the Search for ExtraTerrestrial Intelligence, was this author's #1 reason for his interest in the Moon Farside Protection. It all started in the 1990s, when the French radio astronomer Jean Heidmann (1923–2000) first raised the question of the Moon Farside protection within the activities of the International Academy of Astronautics. In 1999 Heidmann had appointed Maccone his deputy to conduct the IAA "Cosmic Study 1.6" about the relevant both scientific and legal problems. But Heidmann had untimely passed away on July 2nd, 2000, and so Maccone has been in charge of facing all these problems on behalf of the IAA ever since.

NASA did about a year of SETI searches in between October 12, 1992 and October 3, 1993, but was forbidden by Congress to do any SETI activity in the next 25 years (though Astrobiology was supported by NASA in the meantime). Surprisingly, NASA started again to take an interest in SETI in 2018. On September 26–28, 2018, the first NASA Technosignatures Workshop was held in Houston at the Lunar and Planetary Science (LPI) Institute, website https://www.hou.usra.edu/

meetings/technosignatures2018/agenda/. Maccone was invited to make a presentation about "SETI in Europe" and he did so, adding the Moon Farside Protection, see https://www.hou.usra.edu/meetings/technosignatures2018/presentation/?video=maccone.mp4.

8.7 Past and Present Studies About the Moon Farside Protection

There were several Studies of the value of the Moon's far side for various science applications, including radio astronomy. They indicate the place of the current study in the overall "back to the Moon" activity. In particular:

(1) A useful start here is the following paper (and a corresponding ESA M4 proposal): Mimoun et al. 2011, Exp. Astr., https://doi.org/10.1007/s10686-011-9252-3.
(2) The IAA Cosmic Study 1.6, entitled "Protected Antipode Circle on Lunar Farside", was edited by this author with the support of several technical expert with the IAA and legal experts with the IISL (International Institute of Space Law). This IAA Cosmic Study is currently (December 2018) in press.

8.8 Summary About This author's Work to Legally Protecting Radio Astronomy on the Farside and Within the Quiet Cone in the Space Above the Farside

The goal of this paper was to make the readers sensitive to the importance of protecting the Central Farside of the Moon from any future wild, anti-scientific exploitation. The Farside of the Moon is a unique place for us in the whole universe: it is close to the Earth, but protected from the radio emissions that we ourselves are creating in an ever increasing amount, making our radio telescopes blinder and blinder.

In particular, we gave sound scientific reasons why the PAC, Protected Antipode Circle, should be declared an internationally protected area under the Protection of the United Nations, or, in absence of that institution, by direct agreement among the space-faring nations. To achieve this goal, in the mentioned IAA Cosmic Study we analysed the given legal situation and presented a procedure for establishing the PAC. This would encompass the declaration of the PAC as 'international scientific preserves' or 'scientific reserves', although there is a possibility of ramification on the level of domestic law and regulation. This effort would require a broad constituency of nations and could be pursued based on Art. 7(3) of the Moon Agreement.

Furthermore, we showed that the existing Radio Regulations could guarantee the necessary communication frequencies for a future base on the lunar Farside, as well as the necessary radio silence for successful astronomic endeavours within the PAC. We also summed up several ideas for the scientific use of the Farside to gain further knowledge of the universe (and its inhabitants, i.e. SETI).

The Farside cannot be left to the realtor's speculations! And this is an urgent matter!

Some international agreement must be taken for the benefit of all humankind.

8.9 Conclusions: The Moon Village Should Be Located Outside the PAC and Along the 180 Degrees Meridian, Possibly Close to the South Pole

We conclude this short review paper by proposing (again, see also [1]) that the new "Moon Village" supported by the vision of the ESA Director General, Jan Woerner, be located OUTSIDE the PAC (obviously not to interfere with the detection of radiation coming from space) but also SOUTH OF THE PAC, to be "close" to the South Pole as much as needed to benefit of water there. It thus appears *the best venue for the "Moon Village" would be on or around the 180 degree meridian and south to the* $-30°$ *in latitude of the PAC, possibly much more south of that, almost at the South Pole, thus resolving Moon Village VENUE ISSUE.*

References

1. C. Maccone, Moon Farside Protection, Moon Village and PAC (Protected Antipode Circle). Acta Astronautica **154**, 233–237 (2019)
2. C. Maccone, The Quiet Cone above the Farside of the Moon. Acta Astronautica **53**, 65–70 (2003)
3. C. Maccone, Moon Farside Radio Lab. Acta Astronautica **56**, 629–639 (2005)
4. S. Pluchino, N. Antonietti, C. Maccone, in *Protecting the Moon Farside Radiotelescopes from RFI produced at future Lagrangian-Points Space Stations*, paper IAC-06-D4.1.01 presented at the International Astronautical Congress held in Valencia (Spain), 2–6 October 2006
5. C. Maccone, Protected antipode circle on the Farside of the Moon. Acta Astronautica **63**(1–4), 110–118 (2008)

Chapter 9
Exobiolab: Different Life on Different Planets

Sabrina Masiero and Marco Sergio Erculiani

9.1 Introduction

If at first the idea is not absurd then there is no hope for it.
Albert Einstein

The search for extraterrestrial life is undoubtedly one of the most attractive topics in science, particularly to children. With increasing evidence to suggest that the majority of Sun-like stars play host to their own planetary systems, the idea of alien life is seeming ever-more realistic. When we talk about extraterrestrial life, the prevalent popular conception is the green aliens with two big eyes, two legs and two arms. In other words, life with anthropomorphic characteristics. We know well it takes hard work to counter this and other misconceptions, particularly if we aim to remove them from the minds of children. The idea about a laboratory on exoplanets and alien environments may seem at first a tad wild, like the one that animated—and led—to the birth of the Center GAL Hassin in Isnello. The first idea to create an ExoBioLaboratory was born some years ago to involve young students in developing their imagination and curiosity about life on other worlds. The ability to imagine is what drives creativity, enables clear thinking and inspires a sense of humanity. The search for life has had a new enthusiastic restart in the last two decades thanks to the large number of new worlds discovered. The about 4200 exoplanets found so far, show a large diversity of planets, from hot giants to rocky planets orbiting small and cold

S. Masiero (✉)
GAL Hassin Foundation - International Center for Astronomical Sciences, Isnello (PA), Italy
e-mail: sabrina.masiero@inaf.it

Palermo Astronomical Observatory, Italian National Institute for Astrophysics (INAF), Palermo, Italy

M. S. Erculiani
IRA-Istituto di Radioastronomia, Italian National Institute for Astrophysics (INAF), Bologna, Italy
e-mail: marco.erculiani@inaf.it

S. Montebugnoli and A. Melis (eds.), *The Search for ExtraTerrestrial Intelligence*, Springer Proceedings in Physics 260, https://doi.org/10.1007/978-3-030-63806-1_9

stars. Most of them are very different from those of the Solar System and one of the striking case is that of the super-Earths, rocky planets with masses ranging between 1 and 10 $M_{\oplus}$ with dimensions up to twice those of Earth. The long list motivates many scientists to consider which of these worlds could support life and what type of life could live there. One of the most fascinating questions is probably what color alien plants would be. The question matters scientifically because the surface color of a planet can reveal whether anything lives there—specifically, whether organisms collect energy from the parent star by the process of photosynthesis. Photosynthesis is adapted to the spectrum of light that reaches organisms. This spectrum is the result of the parent star's radiation spectrum, combined with the filtering effects of the planet's atmosphere and, for aquatic creatures, of liquid water. Light of any color from deep violet through the near-infrared could power photosynthesis. Around stars hotter and bluer than our Sun, plants would tend to absorb blue light and could look green to yellow to red. Around cooler stars such as red dwarfs, planets receive less visible light, so plants might try to absorb as much of it as possible, making them look black. The ExoBioLaboratory aims to promote the student's curiosity about this area of research, and to make it more attractive through hands-on learning. As students now become protagonists, the motivation therefore appears for them to shed the passivity and estrangement with which they often react to face-to-face lessons. The main activity is to recreate a plausible exoplanetary environment—imagining a Super-Earth orbiting around a star and considering what kind of plants could be born there, and generally speaking, what types of plant life could evolve on such a planet. By educating children about the life-essential conditions here on Earth and comparing them to those of other worlds, they learn that this is the only place in the universe that is suitable for life as we know it, which promotes respect for the environment and a sense of a global community.

9.2 Background Information

9.2.1 Ingredients to Sustain Life

The most important ingredient to sustain life as we know it is liquid water. The presence of liquid water depends on environmental conditions like air temperature and atmospheric pressure. The main driver of the surface temperatures of planets is their distance from the central star they orbit. The temperatures are just right only in a small window so that water does not completely evaporate or freeze. These conditions are modified by local influences like the density of the atmosphere and the composition of potential greenhouse gases. This defines a range around a given star in which liquid water could be present. This range is defined as the "habitable zone". If a planet is found orbiting in this zone, it may potentially possess water in the liquid form and thus sustain life as we know it. In the Solar System, Earth occupies the habitable zone. There is no guarantee that any planet orbiting within the habitable zone actually possesses notable amounts of liquid water or harbours life, because the conditions on any given planet can be very different. Other boundary conditions that may help to sustain life are energy sources (light, chemical) and magnetic fields to protect from ionising particle radiation.

9.2.2 *Life on Exoplanets*

We have no precise information on the structure of the surfaces of extrasolar planets. Still, we have some hints on the chemical composition of their atmospheres, and on the internal chemical composition of their interiors (if they are rocky or fluffy, that is). It is also possible to use the Earth as a model in order to estimate its ecological limits to life and understand, should approximately Earth-like planets be found, whether these parameters could apply. Of course, this does not mean that there is life, but that the possibility could exist for them to host Earth-like lifeforms. This way, (Table 9.1) we can isolate some constraints that limits the main physical parameters span.

Table 9.1 Limits of life. *Credit* McKay, C. P., Requirements and limits for life in the context of exoplanets, 2014 [1]

Parameter	Limit	Note
Low temperature	$\sim -15\,°C$	Limited by liquid water associated with thin films or saline solutions
Upper temperature	122 °C	Solubility of lipids in water, protein stability
Maximum pressure	1100 atm	
Low light	$\sim 0.01\ \mu mol\ m^{-2} s^{-1} = 5 \times 10^{-6}$ direct sunlight	Algae under ice and deep sea
pH	0–12.5	
Salinity	Saturated NaCl	Depends on the salts and temperature
Water activity	0.6	Yeasts and molds
	0.6	Bacteria
UV	$\geq 1000\ J\ m^{-2}$	D. radiodurans
Radiation	50 Gy/h	D. radiodurans growt with continuous dose
	12,000 Gy/h	Acute dose, higher for dry or frozen state[1]

9.2.3 Ecological Limit to Life

As described in McKay [1], temperature is key both because of its influence on liquid water and because it can be directly estimated from orbital and climate models of exoplanetary systems. Life can grow and reproduce at temperatures as low as −15°C, and as high as 122 °C. Studies of life in extreme deserts show that on a dry world, even a small amount of rain, fog, snow, and even atmospheric humidity can be adequate for photosynthetic production producing a small but detectable microbial community. Life is able to use light at levels less than 10^{-5} of the solar flux at Earth. UV or ionizing radiation can be tolerated by many microorganisms at very high levels and is unlikely to be life limiting on an exoplanet. Biologically available nitrogen may limit habitability. Levels of O_2 over a few percent on an exoplanet would be consistent with the presence of multicellular organisms and high levels of O_2 on Earth-like worlds indicate oxygenic photosynthesis. Other factors such as pH and salinity are likely to vary and not limit life over an entire planet or moon.

9.2.4 Mass Versus Gravity

In order to understand if exoplanets could support life, we can understand what gravity acceleration could experience an organism on it. From the physics we know that the superficial gravity depends on the mass m and radius $\mathbf{r}$ according to Newton's formula $\mathbf{a}(\mathbf{r}) = -Gm\hat{\mathbf{r}}/\mid \mathbf{r} \mid^2$, with G the universal gravitational constant. It follows that planets with different masses and sizes would have different gravity values on their surfaces. In [2] a considerable number of the extrasolar planets have a surface gravity very similar to that of Earth. It is true that in the Solar System the mass of rocky planets smaller than Venus grows with the square root of the mass while for gaseous giant exoplanets, the surface gravity linearly grows with the mass. For planets between 1 and 100 Earth masses the surface gravity is roughly similar to that of Earth.

In the ExoBioLaboratory a Super-Earth scenario will be considered for four different stars, the surface gravity being about 1 g (the gravity on the Earth) (Fig. 9.1).

9.2.5 Gravitropism

Gravitropism is the movement or growth of plants in response to gravity. Depending on the gravity of an exoplanet, we can expect different height of plants, if present there. On Earth the height limit is represented by the remarkable 130 m of sequoias. Beyond this limit, gravity can create crucial mechanical damage to the plants. In mosses, hypergravity up to 10 g can increase the size of chloroplasts, photosynthesis and growth, particularly in Specimens of Physcomitrella patens [3].

Hypergravity not only suppresses the growth of stretching, but promotes lateral expansion of plant organs. Plant organs are highly resistant to gravitational acceleration and growth parameters vary in proportion to the logarithm of the magnitude of gravitational acceleration up to 300 g. In [4] we see how in microgravity conditions, the intrinsic structural and physiological anisotropies of plant organs could cause a spontaneous automorphic curvature, while in the presence of gravity, the organs are forced to grow along the vector of gravity, leading to gravitropic curvature. Let us take into account too [5] to become familiar with fundamental aspects of the microcravity environment of space and with plants on Earth (pigment, chlorophyll, different species of plants).

9.2.6 Biosignatures

In the right environment, Super-Earths could be the cradle of alien life that could modify the chemical composition of their atmospheres. So, the search for life signatures requires as the first step the knowledge of planet atmospheres. Indeed, the quest for the determination of the chemical composition of those planetary atmospheres rises also more general interest than that given by the mere directory of the atmospheric compounds. It opens out to the more general speculation on what such detection might tell us about the presence of life on those planets. As, for now, we have only one example of life in the universe, we are bound to study terrestrial organisms to assess possibilities of life on other planets and guide our search for possible

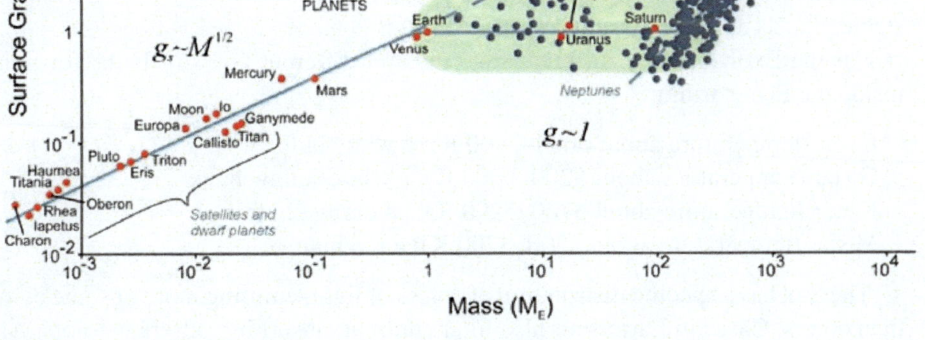

Fig. 9.1 Mass versus surface gravity. The surface gravity is normalized with terrestrial gravity [2]

extinct or extant life on other planetary bodies. Plant-life absorbs visible light, using photosynthesis to convert that light into energy. Visible light is more likely to bounce around within the leaf to give it the best chance of absorbing the useful wavelengths, whereas infrared light is more likely to be reflected and transmitted. (More green light is reflected than other visible wavelengths, but the variation is no more than a few percent.) Leaves are comprised of water-filled cells surrounded by air and the structure of the cell causes some light to be internally scattered, which either exits the leaf as transmitted light below the leaf, or reflected light above the leaf. The reflected infrared light creates a sharp peak in the spectrum known as the vegetation red edge. It is still not fully understood why plants reflect infrared light, but some studies suggest that it is to avoid damage by overheating. It is possible that the detection of the vegetation red edge on other planets would serve as evidence of life.

9.3 The ExobioLaboratory

The ExoBioLaboratory will be developed at the Planetarium of GAL Hassin Foundation—International Center for Astronomical Sciences at Isnello, Sicily (Italy). The Planetarium setting lends itself well as an imaginary planetary environment thanks to the possibility of exploiting the colored lights of the Planetarium and the space available to recreate–with plants and other effects—a hypothetical exoplanetary environment. Another space available at the GAL Hassin Center can be the Solar Laboratory, a dark windowless room where it is possible to build a science laboratory. The ExoBioLab is targeted to Italian students of Secondary School First Grade (aged 11–14) and Secondary School Second Grade (aged 15–16).

The basic idea is to realize 4 different environments for 4 different kinds of exoplanets, according to the 4 different types of stars: F, G, K and M stars. On Earth the processes that led to the birth of life took about one billion years. For this reason, the search for extrasolar planets—and even more the search for life—is directed towards rocky planets (or Super-Earths) orbiting around four main spectral types of stars: F, G, K and M. The spectral type is directly related to the lifetime of a main sequence star and, consequently, to the possibility of formation of long-lived and stable planets adequately suited to develop life.

For the ExoBioLab, 4 different stars mean 4 different kinds of bulbs for lights inside the Planetarium:

F star (temperature about 6000–7500 K): a white bulb;
G star (temperature about 5200–6000 K): a white-jellow bulb;
K star (temperature about 3700–5200 K): an orange bulb;
M star (temperature about 2300–3700 K): a red bulb.

The Sun has a specific distribution of colors of light, emitting more of some colors than others. Gases in Earth's air also filter sunlight, absorbing different colors. As a result, more red light particles reach Earth's surface than blue or green light particles, so plants use red light for photosynthesis. There is plenty of light for land plants, so they do not need to use extra green light. But not all stars have the same distribution of light colors as our Sun. Photosynthesis on extrasolar planets will not necessarily look the same as on Earth.

The graph of Fig. 9.2 shows the intensity of light by color (wavelength) that reaches the surface of Earth-like planets orbiting different types of stars. From hotter to cooler, the star types are F, G, K, and M. Our Sun is represented by the yellow line (G2 star). A planet orbiting an F2 star (red line) has more blue light at the surface, whereas Earth and the K2 star planet receive more red light. Planets around M stars receive much less visible light but much more infrared light. Atmospheric gases such as ozone (O_3), oxygen (O_2), water vapor (H_2O), and carbon dioxide (CO_2) absorb light at specific wavelengths, producing the pronounced dips that astronomers might someday detect in exoplanets' atmospheres. On diagram's horizontal axis, the colored bar marks the wavelengths from 0 to 0.4 microns as UV, 0.4 to 0.7 as visible, and longer than 0.7 as infrared.

9.3.1 F Star—White Bulb

Main sequence stars brighter than the Sun (spectral types F and A and the very short-lived B and O) emit more blue and ultraviolet light than the Sun. Given sufficient time for Earth-type photosynthetic life to evolve (about hundreds of millions to billions of years), planets around such stars could develop an oxygen atmosphere with a layer of ozone that blocks more energetic but potentially harmful ultraviolet light, but which anyway would transmit more blue light to the ground than on the Earth. In response, life could evolve a type of photosynthesis that strongly absorbs blue light, and probably green as well. In contrast, yellow, orange, and red wavelengths of light would likely be reflected by such plants, so the foliage would have the bright colors found during autumn in Earth's deciduous forests all year round. On the other hand, some plants may reflect some blue light due to its overabundance and potential to "burn" photosynthetic organisms (like sunburn from ultraviolet exposure on Earth).

9.3.2 G Star—White-Yellow Bulb

Type G stars are stars similar to our Sun. The basic model is our Solar System with the Sun as a star. In the process of photosynthesis, plants convert energy from the Sun into chemical energy in the form of glucose, or sugar. The chlorophyll in plants absorbs more blue and red light from sunlight, and less green light. Chlorophyll is green, because it reflects green light more than blue and red one.

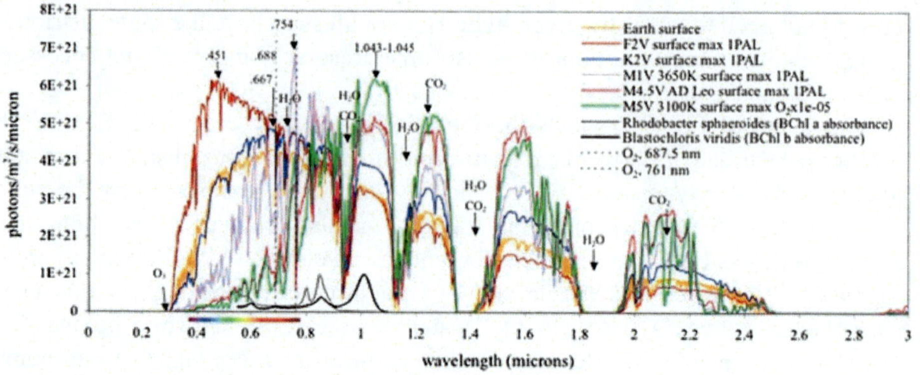

Fig. 9.2 *Credit* NAI—NASA Astrobiology Institute

9.3.3 K Star—Orange Bulb

According to a new study [6], K stars are more likely to host habitable exoplanets than other types of stars. They live a long time, providing surrounding planets plenty of time for the development of life. More importantly, K stars feature less electromagnetic turbulence. The larger, more frequent solar flares produced by M stars can strip away the atmosphere of inner planets, quashing the chance for the development of life.

9.3.4 M Star—Red Bulb

Since M dwarfs comprise ~70% of all stars in the galaxy [7], they offer the best chance of finding habitable planets through sheer numbers and proximity to the Sun. They also offer clear observational advantages. Small planets are easier to detect orbiting small stars via the radial velocity and transit techniques, as spectroscopic Doppler shifts and photometric transit depths are larger due to the smaller star-to-planet mass and size ratios, respectively. Also, because of the relatively low temperatures and luminosities of M dwarfs, their habitable zones are much closer to the stars than those of Sun-like stars, increasing the geometric probability of observing a transit [8], as well as the frequency of transits of habitable-zone planets during a given observational time period. Small planets orbiting small, low-mass stars are also better suited to the application of transmission spectroscopy methods to characterize their atmospheres.

Additionally, the lengthy stellar lifetimes of M-dwarf stars is a benefit. They are extremely long-lived, with main-sequence lifetimes of trillions of years for the lowest-mass M dwarfs. They would therefore offer plentiful timescales for planetary and biological development and evolution on orbiting planets.

However, previous overview papers of nearly a decade ago summarized the current understanding of the instrumental requirements for observing M-dwarf planets, and addressed outstanding questions concerning the impact on surface life of stellar flare activity, synchronous rotation, and the likelihood of photosynthesis on M-dwarf planets [9]. At the time of these papers, no terrestrial-sized planets had yet been discovered around M-dwarf stars, and no statistical information about the occurrence rate of exoplanets as a function of stellar type was available.

9.3.5 Plants on Exoplanets

In the Planetarium it will be possible to create various exoplanetary environments with different kinds of plants, and lighting meant to simulate the light of distant stars as filtered by the extrasolar atmosphere. It is not yet known what the surface characteristics of such exoplanets are. Earth-like environments are assumed here. On Earth there are plants of different colors: in very bright areas the photosynthetic pigments are secreted by the plants in order to form a photosynthetic pigment which is a sort of protective cream or protective filter to shield strongly energetic radiation. In low light conditions, instead, the pigmentation tends to be a darker, color, to collect a greater amount of the radiation that arrives. Light of any color from deep violet through the near-infrared could power photosynthesis. Around stars hotter and bluer than our Sun (F stars), plants would tend to absorb blue light and could look green to yellow to red. In dim habitats, alien vegetation would need more photosynthetic pigments that capture radiation in a wider range of wavelengths, which would give them a dark appearance like many dark plants and flowers on Earth. Plants on planets orbiting dim stars or in distant orbits around brighter stars may appear dark because they need to capture more of the visual and even of the nearer parts of the infrared and UV spectrum. The following types of trees and plants that do live on Earth are suggested as a blueprint of sorts for possible extraterrestrial plants.

9.3.5.1 Plants on Exoplanets Around F Stars

Ginko Biloba (tree) initially green, fan-shaped leaves take on a brilliant yellow colour in autumn.

Liquidambar (tree): the rich dark green, smooth, shiny, star-shaped leaves generally turn brilliant orange, red, and purple colors in the autumn.

9.3.5.2 Plants on Exoplanets Around G Stars

Planet earth and the sun are the most natural system to consider. At the same time, life on Earth will be the reference point for rocky planets or super-earths around G stars. In this case, the green plants most commonly found in our country will be the ones taken into account.

9.3.5.3 Plants on Exoplanets Around K Stars

Festuca Glauca (Blue fescue grass) is a colorful ornamental grass with icy blue foliage and pale yellow flowers.

9.3.5.4 Plants on Exoplanets Around M Stars

Aeonium arboreum, Ophiopogon Planiscapus and Colacasia Esculenta are four species with dark foliage. A useful reference here is the work of Prof. Rosario Schicchi, Director of the Palermo Botanical Garden and professor at Palermo University. His suggestions and a direct comparison with Prof. Schicchi's criteria should be instrumental not only on the choice of the most suitable plants for the various environments, but also for his study of some types of plants that live in extreme conditions, so much so that they can be selected based on the conditions of temperature, pressure and mass of the planet where they could live.

Aeonium arboreum var. atropurpureum: this beautiful species has large heads of deep purple foliage truly stand out against other greenery. Aeonium arboreum should be used for exoplanets around M stars.

For more helpful ideas on the issue of exoplanets and plants on new worlds we recommend [10].

9.4 Other Questions

The ExoBioLaboratory could be developed in various ways, taking into account some astronomical parameters of the two bodies, star and planet. Students should analyse some of the following questions:

a. Can life exist on a tidally-locked planet?
b. A rocky planet much bigger than Earth, where there is more space on the surface and where gravity is stronger. What happens?
c. What happens on a much smaller rocky planet, where you would have less space on the surface and weigh very little?
d. What happens on a planet that is always cloudy, and there is never a clear day or night?
c. What happens on a planet with a very eccentric orbit?

9.4.1 The Future of Our Earth: A New Planet

9.4.1.1 Anthropocene a New Epoch

Since the last ice age, Earth's climate had been relatively warm and stable. It is clear that our climate is no longer stable and is beginning to warm rapidly. Scientists now agree that human activity, rather than any natural progress, is the primary cause of the accelerated global warming. Agriculture, urbanisation, deforestation and pollution have caused extraordinary changes on Earth [11].

Our actions have driven Earth into a new geological epoch, the Anthropocene. For the first time in our home planet's 4.5-billion year history a single species is dictating Earth's future. Our planet Earth is becoming a new planet. How will be our Earth?

A second laboratory can be developed to study the future of our planet.

An excellent report for students will be [12].

9.4.2 Earth's Primordial Atmophere in Comparison to Other Exoplanets' Atmosphere

Earth can be a model for detecting vegetation on exoplanets. Over the last 500 million years, Earth's surface has changed dramatically, from being ice-covered to having huge forests spread out over land. For most of our home planet's early history, land plants did not exist, but around 465 million years ago, the plants colonized the land. As plants evolved on Earth, the vegetation signal that reveals their presence became stronger.

9.5 Conclusion

The students should be able to:

- explain, in their own words, the phenomenon observed;
- record and analyse data on their own and draw the necessary conclusions;
- understand the key conditions needed to find potential life outside Earth;
- create, imagine (and live for a while) a new planet very different from Earth;
- know Earth's natural history as a guide to spot vegetation on exoplanets;
- understand the Earth is a finite sphere with finite resources that can be depleted by mankind. The Earth's atmosphere is very thin compared to its diameter. Humans can easily alter the composition of this thin atmosphere. If too much greenhouse gasses are put into the atmosphere, the Earth will warm because of a stronger greenhouse effect. This has dramatic consequences for our civilization, such as rising sea level, wider deserts, altering climates, and a runaway warming effect to increase the global temperature even more. With no known alien life to help us, or closeby habitable planets, we depent on the Earth;
- learn that this is the only place in the universe that is suitable for life as we know it, which promotes respect for the environment and a sense of a global community.

References

1. C.P. McKay, Requirements and limits for life in the context of exoplanets. PNAS **111**(35), 12628–12633 (2014)
2. F.J. Ballesteros, B. Luque, Walking on exoplanets: is star wars right? Astrobiology **16**(5), 325 (2016)
3. K. Takemura, H. Kamachi, A. Kume, T. Fujita, I. Karahara, Y.T. Hanba, A hypergravity environment increases chloroplast size, photosynthesis, and plant growth in the moss physcomitrella patens. J. Plant Res. **130**, 181–192 (2017)
4. T. Hoson, Plant growth and morphogenesis under different gravity conditions: relevance to plant life in space. Publ. Med. **4**(2), 205–16 (2014)

5. G.L. Vogt, N.P. Moreno, S. Countryman, Plants in space. activity: how does microgravity affect plant growth? (BCM—Baylor College of Medicine, 2012)
6. G.N. Arney, The K Dwarf advantage for biosignatures on directly imaged exoplanets. ApJL **873**(L7), 1–9 (2019)
7. J.J. Bochanski, S.L. Hawley, K.R. Covey, et al., The luminosity and mass functions of low-mass stars in the galactic disk. II. Field AJ **139**, 2679 (2010)
8. P. Nutzman, D. Charbonneau, Planetary exploration and science: recent results and advances. Publ. Astron. Soc. Pac. **120**, 317 (2008)
9. J.C. Tarter, P.R. Backus, R.L. Mancinelli et al., A reappraisal of the habitability of planets around M dwarf stars. Astrobiology **7**, 30–65 (2007)
10. N.Y. Kiang, The color of plants on other worlds, in *Green Aliens are So Passé* (Scientific American, 2008)
11. J. Zalasiewicz, M. Williams, A. Haywood, M. Ellis, The Anthropocene: a new epoch of geological time? **369**, 835–841 (2011)
12. N.L. Allison, R.A. Bindoff, P.M. Bindschadler, N. Cox, M.H. de Noblet, J.E. England, N. Francis, A.M. Gruber, D.J. Haywood, G. Karoly, C. Kaser, T.M. Le Quéré, M.E. Lenton, B.I. Mann, A.J. McNeil, S. Pitman, E. Rahmstorf, H.J. Rignot, S.H. Schellnhuber, S.C. Schneider, R.C.J. Sherwood, K. Somerville, E.J. Steffen, M. Steig, A.J. Visbeck, Weaver, *The Copenhagen Diagnosis, Updating the World on the Lastest Climate Science* (The University of New South Wales Climate Change Research Centre (CCRC), Sydney, Australia , 60 pp., 2009)

Chapter 10
Involvement of the Sardinia Radio Telescope in the Breakthrough Listen Initiatives

Andrea Melis, Delphine Perrodin, Nicolò Antonietti, Marta Burgay, Raimondo Concu, Claudio Maccone, Emilio Molinari, Stelio Montebugnoli, Alessandro Navarrini, Pierpaolo Pari, Maura Pilia, and Andrea Possenti

Abstract The Breakthrough Listen (BL) Program is the most comprehensive search for extraterrestrial life ever done. One hundred million dollars have been invested, leading to thousands of hours of dedicated telescope time at two of the major radio astronomical facilities in the world, the Green Bank Telescope and the Parkes Telescope. In this paper, we outline how the Sardinia Radio Telescope is becoming involved in the BL initiatives. We present how we are planning to conduct observations for SETI as well as for pulsar and fast radio burst searches.

10.1 Introduction

The Sardinia Radio Telescope[1] (SRT, see Fig. 10.1) [1, 2] is a radio astronomical facility that is now fully operational. The wide variety of scientific studies that can be conducted at the telescope, both in single-dish mode and simultaneously with other facilities, in Very Long Baseline Interferometry (VLBI) mode, make it quite attractive and include: imaging, spectroscopy, polarimetry, pulsars, fast radio bursts (FRB) searches, space debris tracking, space weather activities and so on.

The search for life in the Universe, and in particular in the Milky Way, has now become, to all effects, decisive and timely. The overwhelming majority of radio astronomical facilities in the world are now involved in answering the question: "Are we alone?", among them single-dish antennas as well as one of the main SKA

[1] http://www.srt.inaf.it/.

A. Melis (✉) · D. Perrodin · M. Burgay · R. Concu · C. Maccone · E. Molinari · A. Navarrini · M. Pilia · A. Possenti
INAF Osservatorio Astronomico di Cagliari, Via della Scienza 5, 09047 Selargius, Italy
e-mail: andrea.melis@inaf.it
URL: http://www.oa-cagliari.inaf.it/

N. Antonietti · S. Montebugnoli · P. Pari
INAF Istituto di Radio Astronomia, Via Gobetti 101, 40129 Bologna, Italy

S. Montebugnoli and A. Melis (eds.), *The Search for ExtraTerrestrial Intelligence*, Springer Proceedings in Physics 260, https://doi.org/10.1007/978-3-030-63806-1_10

Fig. 10.1 Sardinia radio telescope

(Square Kilometre Array[2]) precursors, the MeerKAT[3] array. Driven by this world-wide trend, we have, over the last few years, established a meaningful relation with UC Berkeley, which has included the Italian team's visit at UC Berkeley in 2018 for a week. The purpose of the visit was primarily to discuss, with the team operating the Breakthrough Listen (BL) program,[4] whether a collaboration between INAF and the Breakthrough Foundation was of mutual concern and how to move forward.

The following year, a small delegation of the BL team came to Italy to visit, both to do a little scouting at the SRT site and to discuss technical, scientific and logistical aspects, so that SRT could become part of the BL program. In particular, SRT nicely complements the other radio astronomical facilities involved in BL thanks to its searching capabilities with high-frequency receivers. In this paper, we describe the current status of the SRT for conducting observations at up to 115 GHz, how we are preparing so the telescope can be involved in BL activities, as well as details of the agreement between INAF and the Breakthrough Foundation. Finally, we outline our conclusions and future perspectives.

[2]https://www.skatelescope.org/.

[3]https://www.sarao.ac.za/gallery/meerkat/.

[4]https://breakthroughinitiatives.org/initiative/1.

Fig. 10.2 INAF institutes involved in the PON

10.2 Enhancement of the SRT for the Study of the Universe at High Radio Frequencies

SRT currently has three available receivers: the dual-band L-P band [3] (305–410 MHz; 1300–1800 MHz), C-high band (5.7–7.7 GHz) and K-band (18–26.5 GHz) [4]; the latter has seven feeds. However, the telescope was designed to cover the whole electromagnetic field in the frequency range 0.3–115 GHz. In order to achieve such a spectral coverage, as well as improve the telescope capabilities and allow efficient operations at such high frequencies, INAF participated in a call for proposals for grants aimed to enhance research infrastructures, the so-called PON (National Operative Program). INAF was awarded the national tender and obtained 18.7 million of euros to be spent within 32 months, of which 15% were destined to the other two Italian radio astronomical facilities which are both 32-m across, Medicina[5] and Noto.[6] The PON project is carried out by four INAF institutes: Arcetri/Firenze (Florence), Bologna, Cagliari and Catania; Fig. 10.2 shows the geographical positions of the four aforementioned locations in Italy.

[5]http://www.med.ira.inaf.it/parabola32m.html.

[6]http://www.noto.ira.inaf.it/.

Fig. 10.3 Laboratories at the INAF Cagliari

The project is organized into work packages, which include a new metrological system for optimizing pointing performance at high frequencies, the upgrading of the digital backends which are currently based on the ROACH2[7] boards and which are no longer in production, a new HPC (High Performance Computing) for post-processing the data produced by the new digital backends as well as a new data storage unit, and finally the enhancement of our laboratories (see Fig. 10.3) in order to manage maintenance activities.

In addition to the receivers mentioned before, four new receivers will be provided: a 16-feed W-band (86–115 GHz), a 19-feed Q-band (33–50 GHz), a three-band K-Q-W and, finally, a millimeter chamber operating in the 80–116 GHz frequency range, composed of an array of about 300 independent detectors. All of the receivers have already been commissioned, and they are expected to be installed by the end of 2021, after which (at least) a year of commissioning will be needed, so as to make the new receivers available for the call for proposals of 2023. We aim to exploit these new receivers to do SETI as well.

10.3 Involvement of the SRT in the BL program

Over the last few years, INAF and UC Berkeley have interacted on several occasions, both during the annual International Astronautical Congress (IAC) and during other meetings. The possibility that SRT might be considered as a new facility in the BL program has been repeatedly mentioned, without ever coming to a practical conclusion. The turning point was in 2018, when an Italian team (formed by the majority of the authors of this paper) visited Berkeley for a whole week. It was a very fruitful week, during which we went in-depth into several topics and discussed

[7] https://casper.ssl.berkeley.edu/wiki/ROACH-2_Revision_2.

Fig. 10.4 BL & OAC team at SRT

some aspects that had never been considered before. In 2019, a BL delegation formed by Vishal Gajjar and David Mac Mahon flew to Sardinia and visited the SRT, Fig. 10.4 shows a picture taken during that visit.

One of the key science objectives of the BL program is to do extensive searches towards the Galactic center, including at high frequencies. The conventional telescope used for this purpose is the Green Bank Telescope[8] (GBT), which however loses efficiency in case of poor weather conditions at K-band. SRT has to deal with the same issue, but is ideally placed with respect to the GBT for carrying out complementary observations. The key points of the discussion/agreement were as follows:

- Use dedicated time at the Sardinia Radio Telescope to conduct targeted searches;
- In particular, use the SRT's unique high energy capabilities to conduct targeted searches of the Galactic Center at high radio frequencies (C-band, K-band). Complementary to what other radio telescopes can offer;
- Breakthrough Listen observations involve the recording of baseband data. New hardware will be installed at SRT for this purpose;
- In parallel with searching for ET, the data will be used to search for pulsars and FRBs in the Galactic Center.

[8]https://greenbankobservatory.org/science/telescopes/gbt/.

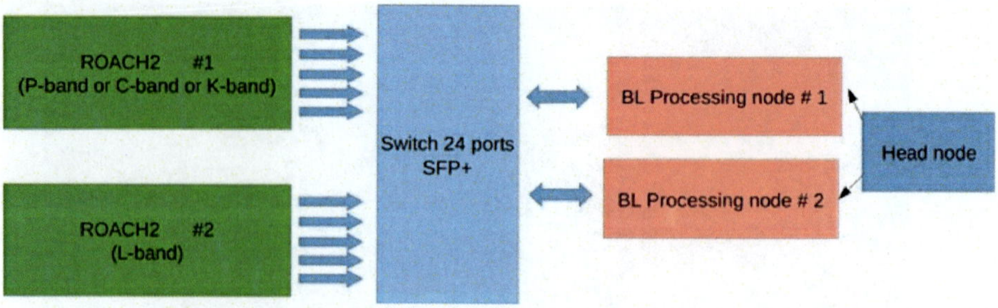

Fig. 10.5 Block diagram of the BL infrastructure

Table 10.1 BL computing node

Chassis	SuperStorage 6049P-E1CR24H 4U
Motherboard	Supermicro X11DPH-T
CPU	2 Intel Xeon Silver 4110 Processor LGA3647
GPU	2 NVIDIA GeForce RTX 3080
NIC	Mellanox MCX4121A-XCAT ConnectX-4 2x10G,
Memory	96 GB DDR4
Hard Disk	24 Toshiba MD04ACA600/HDETS10 6 TB

Figure 10.5 shows the block diagram representing the processing system we are designing in cooperation with BL engineers. We are re-using part of the SARDARA (SArdinia Roach2-based Digital Architecture for Radio Astronomy) [5] backend, both for dedicated SETI observations or as a commensal backend during conventional observing programs of accepted proposals. Two powerful workstations, whose main characteristics are shown in Table 10.1, will be provided by the BL Foundation and will be able to process up to 6 Gbit/s each for searching ET, pulsars and FRBs simultaneously. A detailed description of the hardware and software pipeline used for collection, reduction, archival, and public dissemination of the BL data can be found here [6]; the data collected with the SRT will substantially be processed in the same manner. In a general way, a more comprehensive overview of the facilities involved in the BL program, among which the SRT, can be found here [7].

Finally, during the IAC 2019 held in Washington, the BL's leader Pete Worden announced[9] a new collaboration with scientists working on NASA's Transiting Exoplanet Survey Satellite (TESS), which is shown in Fig. 10.6. The program envisages a survey of the 1,000,000 closest stars to Earth, focusing more towards the center of the Milky Way; moreover, the search is extended to the 100 closest galaxies to ours. SRT will be part of this exciting program; around 12 TESS targets will be observed[10] across 18–26 GHz.

[9] https://breakthroughinitiatives.org/news/27.
[10] https://seti.berkeley.edu/tess/obsplan.pdf.

Fig. 10.6 Artist concept of the transiting exoplanet survey satellite and its 4 telescopes. Credit: NASA/MIT

10.4 Conclusion and Future Remarks

In this paper, we outlined the overview of the ongoing collaboration between INAF and the Breakthrough Listen team. The common aim is for SRT to join the BL activities as soon as possible and to start to do SETI, pulsar and FRB searches in cooperation with all the other facilities involved in the collaboration. In 2020, the installation and commissioning of the described hardware and software infrastructure was supposed to be completed so as to be ready to operate by the end of 2020, however due to the Covid-19 pandemic, the program has been significantly slowed down. Nevertheless, we are confident that, despite these unforeseen circumstances, we will complete all the needed tasks and SRT will soon, finally, start to do SETI in the context of the BL program.

References

1. P. Bolli, et al., Sardinia radio telescope: general description, technical & commissioning and first light. J. Astron. Instrum. **4**(3 & 4), 1550008 (2015). https://doi.org/10.1142/S2251171715500087
2. I. Prandoni, et al., The sardinia radio telescope . From a technological project to a radio observatory. Astron. Astrophys. **608**, id.A40 (2017). https://doi.org/10.1051/0004-6361/201630243

3. G. Valente, et al., The dual-band LP feed system for the Sardinia Radio Telescope prime focus, in *Proceedings SPIE Millimeter, Submillimeter, and Far-Infrared Detectors and Instrumentation for Astronomy V*, vol. 7741, 774126. https://doi.org/10.1117/12.857306
4. G. Valente, et al., Status of the radio receiver system of the Sardina radio telescope, in *Proceedings of the SPIE*, vol. 9914, id. 991425, 13 pp. (2016). https://doi.org/10.1117/12.2232880
5. A. Melis et al., SArdinia roach2-based digital architecture for radio astronomy. J. Astron. Instrum. **7**(1), 1850004 (2018). https://doi.org/10.1142/S2251171718500046
6. M. Lebovsky et al., The breakthrough listen search for intelligent life: public data, formats, reduction, and archiving. Publ. Astron. Soc. Pacific **131**(1006), 124505 (2019). https://doi.org/10.1088/1538-3873/ab3e82
7. V. Gajjar, et al., The breakthrough listen search for extraterrestrial intelligence. Astro2020: decadal survey on astronomy and astrophysics, APC white papers, no. 223. Bulletin Am. Astron. Soc. **51**(7), id. 223 (2019). https://doi.org/10.1088/1538-3873/ab3e82

Chapter 11
SETI Program at the Medicina INAF Radioastronomy Station: Past, Present, Future

Stelio Montebugnoli

Abstract In these last year, new Earth-like type exoplanets have been discovered. These could be suitable to sustain life and maybe able to evolve in a technological civilization as happened on the earth. SETI program for 40 years is looking to verify such an existence through observations in both the radio and optical bands. At the Medicina radio telescope, the SETI program started in the middle of the 90s, exploiting a high-resolution spectrum analyzer, with many million channels, given by the University of Berkeley. We observed in a piggyback mode for ten years, in the period 1998–2008, for 73.000 h of operation. During these observations, we searched for a "Doppler-shifted" monochromatic radio because we thought that ET would have sent such a signal to flag his presence into space. Many suspect signals were received, but no one has been verified in further observation as requested by the Post Detection International Protocol. From 2008 till now, the SETI program didn't continue, but research activities will probably start soon with new post-processing algorithms implemented on a commercial fast Personal Computer.

11.1 Activities at the Medicina Radiotelescopes

11.1.1 Past

SETI program activities started in 1998 at the Medicina radiotelescopes (see Fig. 11.1), at that time belonging to the National Research Council (CNR), and from 2005 on to the National Institute of Astrophysics. Up to now, it was supposed that ET intentionally transmits a monochromatic signal to flag out his presence somewhere in the Milky Way. Such a monochromatic "footprint" should be easily recognizable

S. Montebugnoli (✉)
Associate Senior Engineer Cross Radiotelescope, Medicina, Italia
e-mail: s.montebugnoli@ira.inaf.it

SETI Advisor Scientific Directorate INAF, Rome, Italy

Associate Senior Engineer IRA INAF, Bologna, Italy

S. Montebugnoli and A. Melis (eds.), *The Search for ExtraTerrestrial Intelligence*, Springer Proceedings in Physics 260, https://doi.org/10.1007/978-3-030-63806-1_11

Fig. 11.1 Medicina (Bologna) radiotelescopes

among an ocean of natural radio signals coming to us from the bodies composing the universe. Besides, such a radio carrier is very efficient from an energetic point of view since the overall available power is all concentrated in it. The monochromatic radio signal, by definition, doesn't have any modulation (then no information). Still, In this case, its presence means information on the existence of technological ET. If we wanted to flag out to ET, our presence in the Galaxy ... we would transmit such a signal, of course, this is just a working hypothesis. We are not sure it is this way!

A high-resolution spectrum analyzer performs the extraction of the radio monochromatic signal from the background noise (sky + system). Till 2008, the Serendip IV high-performance spectrum analyzer [1] coming from the Berkeley University (Ca-USA-) laboratories (see Fig. 11.2), has been used. Such a spectrometer had 24 million channels with 16 MHz input bandwidth. Thanks to the SETI Institute (Mountain View, Ca-USA-) support, we had more spectrum analysis boards installed inside the Serendip IV frame, reaching in this way 24 million channels (0.7 Hz as frequency resolution). The Serendip IV system was installed in piggyback mode, in parallel to the ongoing activities, at the 32 mt VLBI dish. In this way, the operational costs were practically reduced to zero, observing for 24 h a day all year long (see Fig. 11.3). Of course, the system didn't work during the antenna maintenance periods. In this configuration, we collected data for about 73.000 h in the radio astronomical bands included in the 1.4–23.0 GHz range. From 2008 on, the

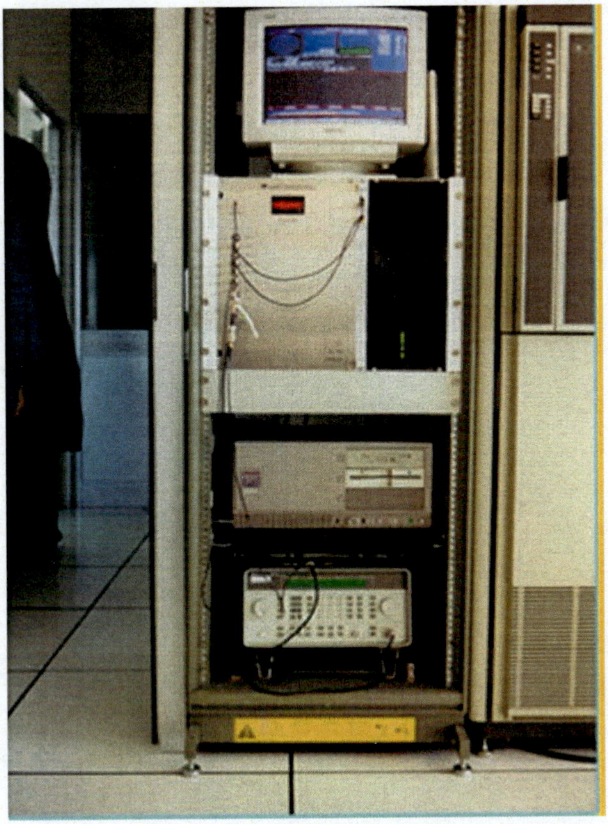

Fig. 11.2 The Serendip IV spectrum analyzer, installed at the 32 m VLBI dish

S IV spectrometer went out of service for some technical failure and obsolescence of some electronic components, not available in the market anymore.

Many suspect signals were received, but unfortunately, no one has never been confirmed in later observation, as required from the IPDP (International Post Detection Protocol). Figure 11.4 reports one of the many received candidates: this was a Doppler-shifted (red/Blue) monochromatic signal as that one we would expect from ET. The antenna was aiming at Mars, and unlikely this was a Rover operating on its surface because the frequency (1421 MHz) is a radio-astronomical band, so not allowed to the Agencies for spacecraft operations. Anyway, we had no confirmation in later observation in the same position, frequency, and time. In Fig. 11.5, another candidate is reported. The radio carrier appears to shift in frequency. Many other intriguing candidate signals have been cataloged while the system was working in piggyback mode at the Medicina VLBI 32 mt dish. Still, no one has never been confirmed as well! Ten years of data were stored on a very capable hard-disk at the Medicina facilities and always available for further data analysis.

It is planned to re-analyze the available Serendip IV data (about 10 GB) using methods different from those used in the first analysis run. These are now under evaluation. Maybe more accurate processing of the data can detect new candidates,

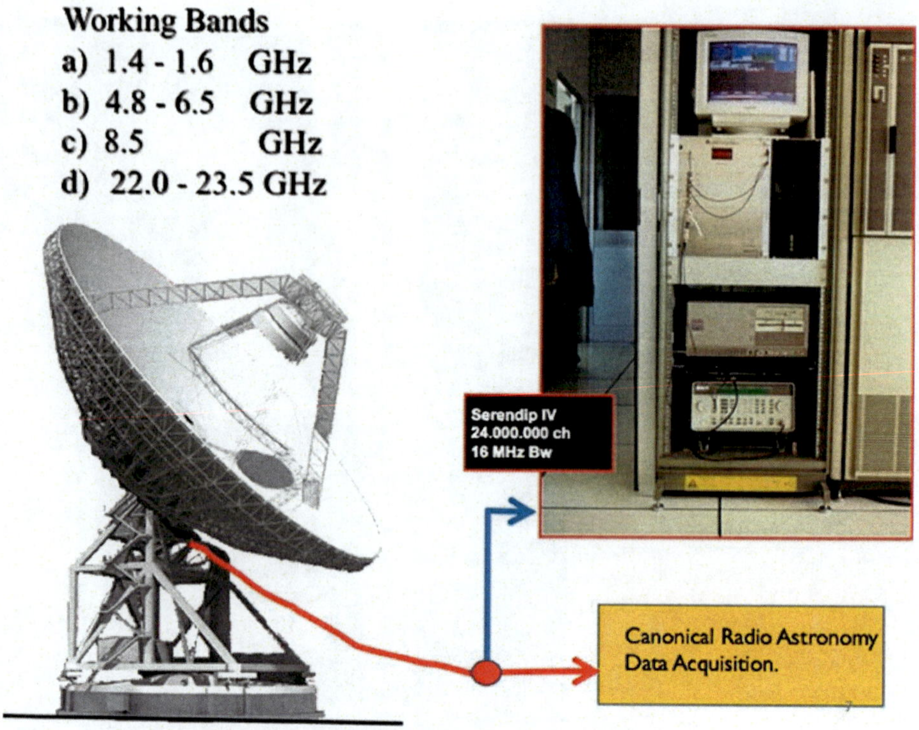

Fig. 11.3 Block diagram scheme of the Serendip IV piggy-back system

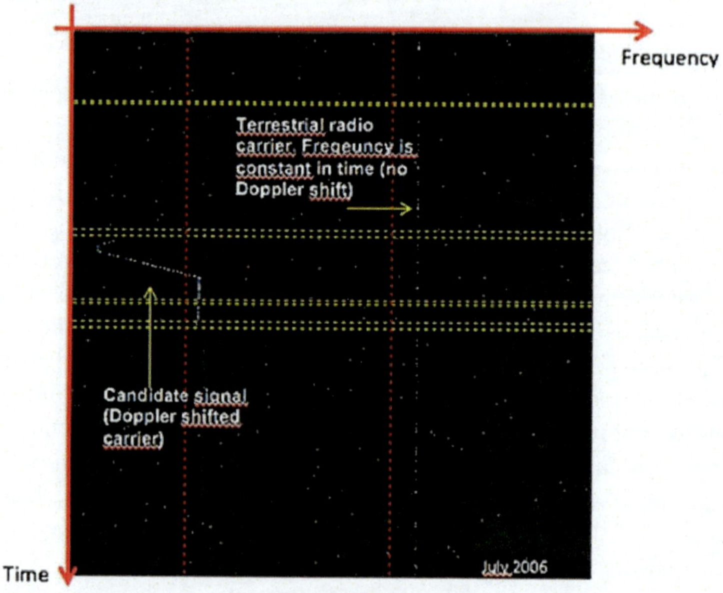

Fig. 11.4 One of the many suspect signals received

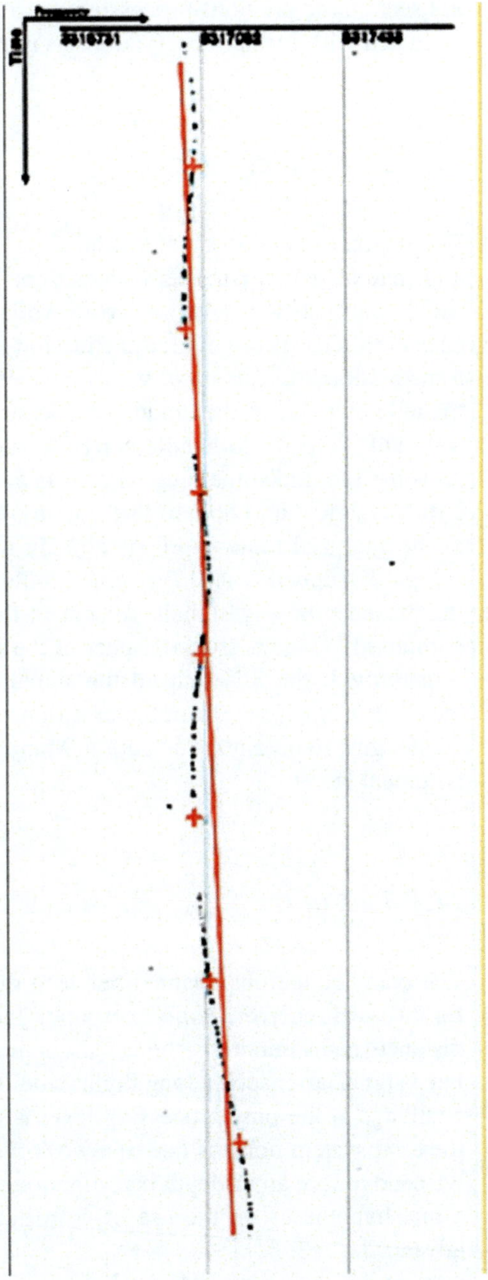

Fig. 11.5 The antenna tracked the same position on the sky for a couple of hours

not seen in the previous processing phase. Of course, if found, these will have to be validated with further observations as requested by the IPDP.

11.1.2 Present

The search for a radio carrier, as an alien signal, didn't give any results till now, then a change of search paradigm needs to be introduced. It is a common opinion in the SETI community it would be worthwhile to search for unintentional signals in the case the aliens spread out radio emission in the Galaxy as part of their planet service transmissions. In this case, we cannot efficiently receive and decode it because of the unknowledge of the modulation scheme. In this case, it is not possible to design a suitable detector to handle correctly such an unknown signal. We could say that "it is hard to find something when it is not clear what we are looking for." We need a mathematical tool able to flag out any weak modulated signal buried in the noise of the analyzed radio band. As already widely demonstrated by *Claudio Maccone* [2], the *Karhunen Loewe Transform* could be very suitable to give information about the presence of any signals present in the band under control; anyhow, these are composed [3, 4]. A crucial aspect of the KLT that needs to be more investigated is sensitivity. If the S/N ratio of the signal to be analyzed is low, the eigenvalues of the eigenvectors, that form the covariance matrix, are close to zero, and it could be challenging to obtain the detection. Many papers have already been published on the argument [5, 6].

11.1.3 Future

The communications channel between the possible alien radio transmitter and the Earth-based receiver (radio telescope) generally is constituted by an unimaginable distance (many hundreds, thousands, or more of light-years). An eventual alien modulated signal, after such a long flight, would reach the receiver antenna as a noise. The challenge at the post-processing level is then how to understand if a small part of the total system noise is due to a weak modulated signal sent by ET or not. Maybe we need to face an in-depth investigation of the possible use of the Entropy concept. Some hypotheses on the use of entropy in the SETI program have already been advanced.

11.2 Some Considerations

Researchers have at their disposal a lot of almost similar mathematical transforms as Fourier, Wavelet, Curvelet, Brushlet, Contourlet, Wave Atom, Haar, and so on, for the post-processing algorithms. Each of these transforms works correctly for specific categories of signals and doesn't work well for other types and presents an N log N computation charge. These math transforms also introduce serious system cache memory problems that arose with handling a vast array of data. The extremely complicated task to verify the presence of an intelligent signal out of the noise leads to a particular approach that could imply the evaluation of the entropy as defined by Shannon [7] and Leidich [8].

11.3 Conclusion

Up to now, no artificial signals from space have shown up at the radiotelescopes output. The common idea is that we have to change the search paradigm. For these reasons, future activities will be oriented to the development of new algorithms as the KLT or other types of math transforms able to detect any modulated weak signal and, at the same time, start to enter into the vast field of the information Entropy evaluation. Generally speaking, we have to change both the concepts of both "how" and "what" to search. It will undoubtedly be a very complex challenge.

References

1. D. Werthimer et al., The Serendip IV Arecibo sky survey—a new era in bioastronomy. Astro. Soc. Pacific Conf. Ser. **213**, 479–483 (2000)
2. C. Maccone, The KLT (Karhunen–Loève transform) to extend SETI searches to broad-band and extremely feeble signals. Acta Astronaut. **67**, 1427–1439 (2010)
3. C. Maccone, Mathematical SETI, 2012. A 724-Pages Book Published by Praxis Springer in the Fall of 2012. ISBN: ISBN-10:3642274366 |ISBN-13:978-3642274367|Edition: 2012
4. S. Montebugnoli, J. Monari, C. Bortolotti, A. Cattani, A. Maccaferri, M. Poloni, P.P. Pari, A. Orlati, S. Righini, S. Poppi, M. Roma, M. Teodorani, D. Caliendo, C. Maccone, C.B. Cosmovici, N. D'Amico, Acta Astronautica **58**, 222–229 (2006)
5. F. Schillirò, S. Pluchino, C. Maccone, S. Montebugnoli, IstitutoNazionale di Astrofisica (INAF)—Istituto di Radioastronomia (IRA) Rapporto Tecnico, La KL Transform: considerazioni generali sulle metodologie di analisi ed impiego nel campo della Radioastronomia, Technical Report (in Italian only), January 2007
6. C. Maccone, Innovative SETI by the KLT, in *Proceedings of the "Bursts, Pulses and Flickering" Conference held at Kerastari, Greece, June 18, 2007 at POS (Proceedings of Science)*, https://pos.sissa.it//archive/conferences/056/034/Dynamic2007_034.pdfS
7. Olivier Rioul LTCI, Telécom ParisTech Université Paris-Saclay 75013 Paris, France CMAP Ecole Polytechniqué Université Paris-Saclay 91128 Palaiseau, France "This is IT: A Primer on Shannon's Entropy and Information" L'Information, Séminaire Poincaré XXIII (2018), pp. 43–77
8. R. Leidich, Introduction to Entropy Transforms, https://agnentropy.blogspot.com. Accessed on 11 May 2017

Chapter 12
Searching for Life on Mars: A Brief Summary

Roberto Orosei

Abstract Mars is today a cold, dry and sterile world with a thin atmosphere made of CO_2. The geologic and compositional record of the surface reveals however that in the past Mars had a thicker atmosphere and liquid water flowing on its surface. For this reason, it has been postulated that life could have developed and that some primitive life forms may be existing even today. This paper will summarize the main discoveries that have led to the current understanding of the geologic, climatic and potentially biologic evolution of Mars, and will provide an overview of current developments and near-future plans for the search for life on the red planet.

12.1 Introduction

Mars is the fourth planet of the Solar System and the second smallest, preceeded only by Mercury. Differentiated in a metallic core and a silicate mantle, Mars has a mass of about one-tenth that of our planet, a diameter slightly less than half the Earth's, and a surface area approximating that of terrestrial landmass. The Martian day is only slightly longer than Earth's, and its rotation axis has a similar inclination.

The Martian surface can be subdivided into three main morphological provinces: the Southern hemisphere consisting of a cratered highland reminiscent of the Moon, the topographically depressed and smooth Borealis basin, covering most of the Northern hemisphere, and the Tharsis plateau, a highland of volcanic origin 5000 km across centered on the equator and encompassing some of the largest volcanoes in the Solar System.

Other major features of the Martian surface are Valles Marineris, the largest canyon in the Solar System with a length of 4000 km and a depth of up to 7 km, and the two impact basins of Hellas Planitia and Argyre Planitia, with a diameter of 2300 km and 1800 km respectively. Mars also possesses two permanent polar ice

R. Orosei (✉)
Istituto Nazionale di Astrofisica, Istituto di Radioastronomia, 40129 Bologna, Italy
e-mail: roberto.orosei@inaf.it

S. Montebugnoli and A. Melis (eds.), *The Search for ExtraTerrestrial Intelligence*, Springer Proceedings in Physics 260, https://doi.org/10.1007/978-3-030-63806-1_12

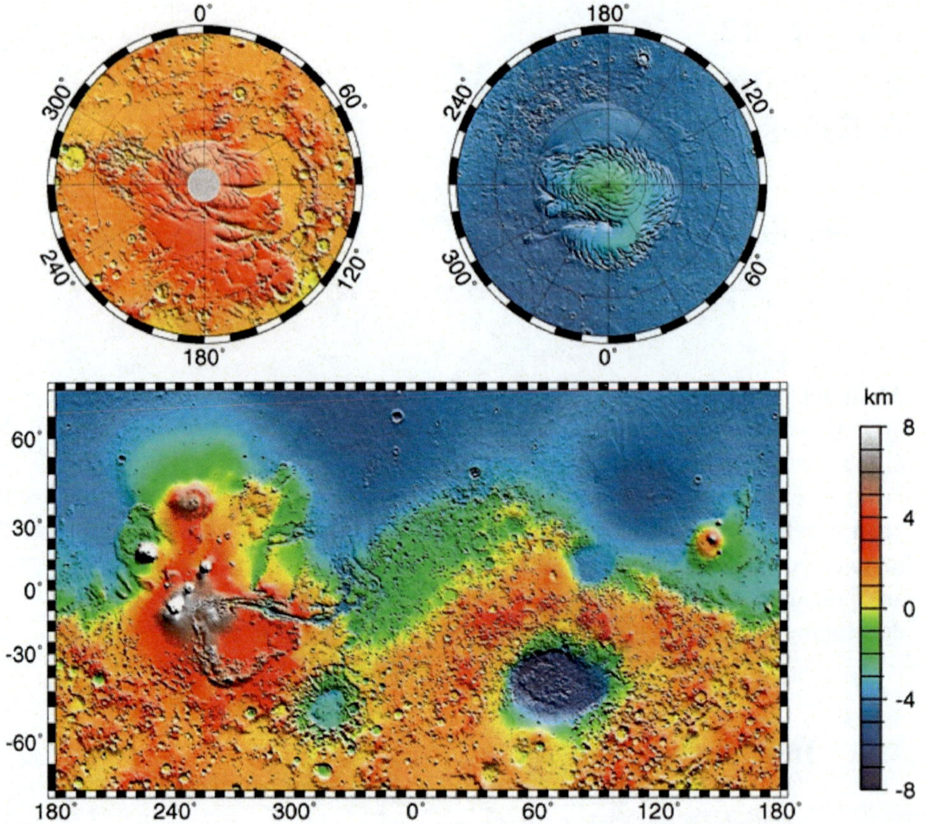

Fig. 12.1 Maps of Mars' global topography. The projections are Mercator to 70° latitude and stereographic at the poles with the south pole at left and north pole at right. The Tharsis volcano-tectonic province is centered near the equator in the longitude range 220° E–300° E and contains the vast east-west trending Valles Marineris canyon system and several major volcanic shields including Olympus Mons (18° N, 225° E), Alba Patera (42° N, 252° E), Ascraeus Mons (12° N, 248° E), Pavonis Mons (0°, 247° E), and Arsia Mons (9° S, 239° E). Major impact basins include Hellas (45° S, 70° E) and Argyre (50° S, 320° E). The map uses an aerocentric coordinate convention with east longitude positive (Image Credit: NASA/JPL/GSFC)

caps, consisting mostly of water ice admixed with some CO_2 and dust, and extending for about 1000 km with a maximum thickness of a few kilometers (Fig. 12.1).

Mars is a terrestrial planet with a rocky surface with a composition corresponding to that of tholeitic basalt, although parts are more silica-rich and may be similar to andesitic rocks [1]. The red-orange colour of the Martian surface is produced by nanophase particles of ferric oxides that constitute the dust covering the entirety of the planet.

12.2 The Exploration of Mars

Since the first detailed images of the Martian surface were acquired by automated probes, in the early seventies of the last century, it became obvious that the planet surface was sculpted by the action of water [2]. Imponent features such as valley networks and outflow channels were already visible in images from Mariner 9. The former are dendritic network of channels resembling a terrestrial fluvial system, while the latter are large stretches of terrain that appear to have been scoured by the flow of huge quantities of water. Other landforms, such as rampart craters and polygonal terrain, hinted at the present of ground ice.

These findings are at odds with the current climate of Mars. The current atmospheric pressure and temperature do not allow the survival of liquid water for any extended period, but the planet does have polar caps that appear to be constituted mostly by water ice [3]. Attempts to reconcile today's conditions on Mars with the evidence of past liquid water on the surface led to the hypothesis that Mars had to have a much denser CO_2 atmosphere in the past that was lost over the ages due to the weak gravitational pull resulting from a mass that is about one-tenth that of the Earth. Such dense atmosphere would produce a greenhouse effect capable of raising the mean surface temperature above the melting point of water [2].

The realization that Mars once had climatic conditions similar to those of the Earth led to the speculation that life could have been developing in the planet's past, and that it could exist even today. As a consequence, NASA launched the two twin Viking spacecrafts, each carrying a lander endowed with a set of experiments meant to detect life through the analysis of Martian soil. Most of the Viking experiments, which were based on the capability of living microorganisms to activate in the presence of liquid water or to use carbohydrates in their metabolism, produced no evidence for life [4], but some still debate whether one of them recorded signs of chemical activity that could be attributed to biotic activity [5]. In hindsight, this outcome could have been expected even if life still survived on Mars because its surface, due to the thin atmosphere, is constantly bombarded by ultraviolet radiation and high energy particles, which would make the survival of even the most tenacious terrestrial microorganisms extremely challenging.

These negative results led to the abandonment of Mars robotic exploration for more than a decade, during which the copious harvest of data collected by the Viking probes was slowly elaborated into a more detailed and consistent picture of the evolution of Mars through the ages. The geologic history of the planet was divided into three main eras, the Noachian, the Hesperian and the Amazonian. The Noachian, lasting a few to several hundred million years, was the age in which Martian climate was more favourable to life, while the Hesperian was characterized by intense volcanic activity, leading to the formation of the Tharsis plateau on which most of the main volcanoes of the planet are located, and by the presence of massive ice sheets. The Amazonian, the modern era of Mars, is the current arid and glacial age of Mars [2] (Fig. 12.2).

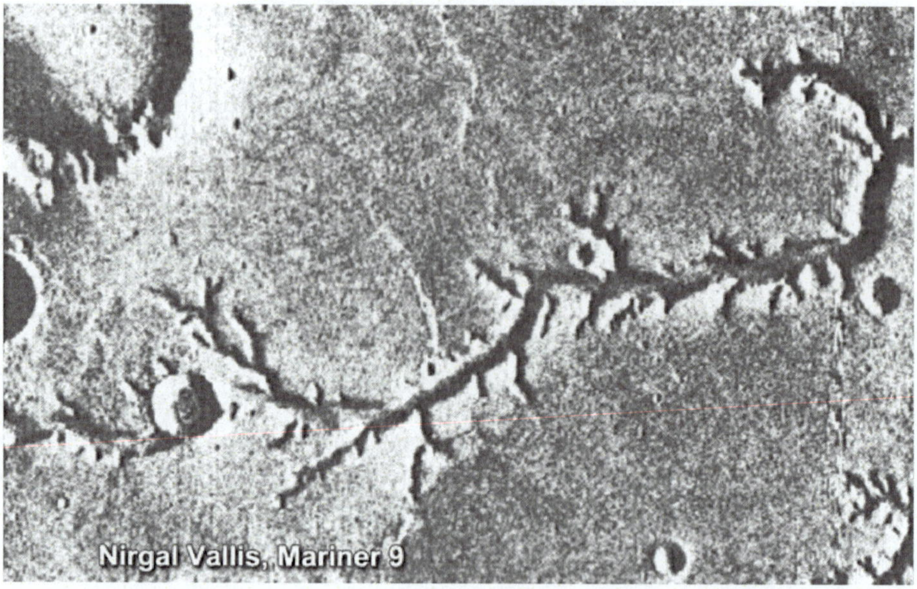

Fig. 12.2 Mariner 9 view of part of Nirgal Vallis. Nirgal Vallis is a 670 km-long east–west trending valley network (Image Credit: NASA/JPL-Caltech)

Several points remained open to debate. The early Sun was perhaps only 70% of its current brightness, and climate models were unable to reproduce sustained periods of warm temperature even in the presence of a thick CO_2 atmosphere (see e.g. [6] for a discussion). Estimates of the quantity of water needed to carve the water-related geologic features observed on the surface of the planet are much greater than the total ice present on the Martian surface [2]. Much uncertainty remained on the onset and decline of volcanic activity of the planet, and on its capability to influence Martian climate and habitability. It was clear, however, that the early Mars presented all factors required for habitability, that is for a living organism to survive, and that such conditions persisted for a time comparable to the one required on Earth for the emergence of life. The study of Mars could thus provide the answer to many questions still lingering on the emergence of life on Earth, and on its likelihood of happening on other planets outside the Solar System. Because of the enormous importance attributed to this scientific endeavour, the exploration of Mars was resumed by NASA in the late eighties of the twentieth century.

12.3 The Search for Life

The initial driver behind the new wave of missions to Mars was summarized by the slogan "Follow the water", because of its importance in understanding the geological, climatic and biological evolution of the planet. As NASA orbiters were followed by landers and rovers, and other space agencies started launching their missions to Mars, more and more evidence was found for an ancient environment in which liquid water persisted on the surface for extended periods of time, the atmosphere was denser than today and a global dipolar magnetic field similar to Earth's protected it from erosion by the solar wind [7]. High resolution images of the surface showed features that were interpreted as due to the recent occurrence of liquid water on the surface [8], although only in limited locales and for brief periods during the warm season. Neutron spectroscopy detected the presence of permafrost in the first meter of soil, extending from the poles to mid-latitudes [9]. Radar sounding allowed the identification of water ice as the predominant constituent of the polar caps [10]. Finally, the measurement of the current erosion rate of the atmosphere due to the solar wind allowed more precise estimates of the total atmospheric loss, including water vapour, over the age of the Solar System [11].

The discovery of methane in the atmosphere of Mars both from the Earth [12] and from Martian orbit [13] started a debate on its origin that is still ongoing. Methane is destroyed by ultraviolet radiation in a relatively short time, and its presence, however small, implies an active source. On Earth, the main processes releasing methane in the atmosphere are volcanism and biologic activity, both of which could not be observed on Mars. A currently operative mission dedicated to the study of trace gases in the Martian atmosphere has yet to provide a definite answer to the question of methane origin, but the recent detection of seasonal variability in the quantity of atmospheric methane seem consistent with sources located at the surface or in the subsurface [14], rather than in the atmosphere.

Another recent development has been the detection of a system of liquid water bodies beneath the south polar cap of Mars [15]. In spite of the theoretical difficulties in reconciling this presence with the very low mean annual temperature at the poles, requiring at a minimum the presence of dissolved salts depressing the freezing point of water [16], and perhaps some thermal anomalies in the crust beneath this area [17], no alternative interpretations have yet been proposed for the radar observations leading to the detection. These subglacial bodies of water constitute the first potential habitat on Mars, being thermally stable and protected from the radiation flooding the surface. However, the lack of any information on the chemical composition and availability of redox pairs in such an environment prevents any assessment of this hypothesis.

Accessing the Martian subsurface to search for evidence of past or present life is considered a high-priority goal in the exploration of the planet, but the technical challenges for such an endeavour are daunting. The Rosalind Franklin rover, to be launched in 2022, will carry a drill capable of reaching a depth of two meters [18], while a recent mission concept for landing a drilling station on Mars foresees a maximum penetration of the order of a hundred meters [19]. Current and planned missions for the search of life on Mars focus on the detection and collection of samples from sedimentary beds or mineral deposits formed in an aqueous environment and accessible on the surface [20].

The detection of life, on Mars and elsewhere, is a complex problem that cannot be solved through a single measurement technique [21]. Because of potential differences between terrestrial and extraterrestrial life, of the relative fragility of complex molecules such as those of biological chemistry, and of the complex alterations induced by a harsh environment such as Mars', the unambiguous identification of life traces will require the simultaneous detection of several biomarkers in the same sample [18]. For this reason, there is a risk that a robotic mission performing in-situ analysis of potential biologic samples will not be able to obtain a definite answer even in the presence of biomarkers. This is why the main goal for Mars exploration in this decade is the return of samples to Earth in pristine conditions [22].

References

1. J.-P. Bibring, Y. Langevin, Mineralogy of the Martian surface from mars express OMEGA observations, in *The Martian Surface—Composition, Mineralogy, and Physical Properties*, ed. by J. Bell III (Cambridge University Press, New York, 2008), pp. 153–168
2. M.H. Carr, *Water on Mars* (Oxford University Press, New York, 1996)
3. S. Byrne, The polar deposits of mars. Annual Rev. Earth Planet. Sci. **37**, 535–560 (2009). https://doi.org/10.1146/annurev.earth.031208.100101
4. H.P. Klein, The viking biological experiments on mars. Icarus **34**, 666–674 (1978). https://doi.org/10.1016/0019-1035(78)90053-2
5. G.V. Levin, P.A. Straat, The case for extant life on mars and its possible detection by the viking labeled release experiment. Astrobiology **16**, 798–810 (2016). https://doi.org/10.1089/ast.2015.1464
6. A.M. Palumbo, J.W. Head, R.D. Wordsworth, Late Noachian Icy Highlands climate model: exploring the possibility of transient melting and fluvial/lacustrine activity through peak annual and seasonal temperatures. Icarus **300**, 261–286 (2018). https://doi.org/10.1016/j.icarus.2017.09.007
7. J.E.P. Connerney, M.H. Acuna, P.J. Wasilewski, N.F. Ness, H. Reme, C. Mazelle, D. Vignes, R.P. Lin, D.L. Mitchell, P.A. Cloutier, Magnetic lineations in the ancient crust of mars. Science **284**, 794–798 (1999). https://doi.org/10.1126/science.284.5415.794
8. M. Malin, K. Edgett, Evidence for recent groundwater seepage and surface runoff on mars. Science **288**, 2330–2335 (2000). https://doi.org/10.1126/science.288.5475.2330
9. W.C. Feldman, W.V. Boynton, R.L. Tokar, T.H. Prettyman, O. Gasnault, S.W. Squyres, R.C. Elphic, D.J. Lawrence, S.L. Lawson, S. Maurice, G.W. McKinney, K.R. Moore, R.C. Reedy, Global distribution of neutrons from mars: results from mars Odyssey. Science **297**, 75–78 (2002). https://doi.org/10.1126/science.1073541
10. J.J. Plaut, G. Picardi, A. Safaeinili, A.B. Ivanov, S.M. Milkovich, A. Cicchetti, W. Kofman, J. Mouginot, W.M. Farrell, R.J. Phillips, S.M. Clifford, A. Frigeri, R. Orosei, C. Federico, I.P. Williams, D.A. Gurnett, E. Nielsen, T. Hagfors, E. Heggy, E.R. Stofan, D. Plettemeier, T.R. Watters, C.J. Leuschen, P. Edenhofer, Subsurface radar sounding of the south polar layered deposits of mars. Science **316**, 92–95 (2007). https://doi.org/10.1126/science.1139672
11. B.M. Jakosky, D. Brain, M. Chaffin, S. Curry, J. Deighan, J. Grebowsky, J. Halekas, F. Leblanc, R. Lillis, J.G. Luhmann, L. Andersson, N. Andre, D. Andrews, D. Baird, D. Baker, J. Bell, M. Benna, D. Bhattacharyya, S. Bougher, C. Bowers, P. Chamberlin, J.-Y. Chaufray, J. Clarke, G. Collinson, M. Combi, J. Connerney, K. Connour, J. Correira, K. Crabb, F. Crary, T. Cravens, M. Crismani, G. Delory, R. Dewey, G. DiBraccio, C. Dong, Y. Dong, P. Dunn, H. Egan, M. Elrod, S. England, F. Eparvier, R. Ergun, A. Eriksson, T. Esman, J. Espley, S. Evans, K. Fallows, X. Fang, M. Fillingim, C. Flynn, A. Fogle, C. Fowler, J. Fox, M. Fujimoto, P. Garnier, Z. Girazian, H. Groeller, J. Gruesbeck, O. Hamil, K.G. Hanley, T. Hara, Y. Harada, J. Hermann, M. Holmberg, G. Holsclaw, S. Houston, S. Inui, S. Jain, R. Jolitz, A. Kotova, T.

Kuroda, D. Larson, Y. Lee, C. Lee, F. Lefevre, C. Lentz, D. Lo, R. Lugo, Y.-J. Ma, P. Mahaffy, M.L. Marquette, Y. Matsumoto, M. Mayyasi, C. Mazelle, W. McClintock, J. McFadden, A. Medvedev, M. Mendillo, K. Meziane, Z. Milby, D. Mitchell, R. Modolo, F. Montmessin, A. Nagy, H. Nakagawa, C. Narvaez, K. Olsen, D. Pawlowski, W. Peterson, A. Rahmati, K. Roeten, N. Romanelli, S. Ruhunusiri, C. Russell, S. Sakai, N. Schneider, K. Seki, R. Sharrar, S. Shaver, D.E. Siskind, M. Slipski, Y. Soobiah, M. Steckiewicz, M.H. Stevens, I. Stewart, A. Stiepen, S. Stone, V. Tenishev, N. Terada, K. Terada, E. Thiemann, R. Tolson, G. Toth, J. Trovato, M. Vogt, T. Weber, P. Withers, S. Xu, R. Yelle, E. Yiğit, R. Zurek, Loss of the Martian atmosphere to space: present-day loss rates determined from MAVEN observations and integrated loss through time. Icarus **315**, 146–157 (2018). https://doi.org/10.1016/j.icarus.2018.05.030

12. M.J. Mumma, R.E. Novak, M.A. DiSanti, B.P. Bonev, A sensitive search for methane on mars. Bullet. Am. Astron. Soc. **35**, 937 (2003)
13. V. Formisano, S. Atreya, T. Encrenaz, N., Ignatiev, M. Giuranna, Detection of methane in the atmosphere of mars. Science **306**, 1758–1761 (2004). https://doi.org/10.1126/science.1101732
14. C.R. Webster, P.R. Mahaffy, S.K. Atreya, J.E. Moores, G.J. Flesch, C. Malespin, C.P. McKay, G. Martinez, C.L. Smith, J. Martin-Torres, J. Gomez-Elvira, M.-P. Zorzano, M.H. Wong, M.G. Trainer, A. Steele, D. Archer, B. Sutter, P.J. Coll, C. Freissinet, P.-Y. Meslin, R.V. Gough, C.H. House, A. Pavlov, J.L. Eigenbrode, D.P. Glavin, J.C. Pearson, D. Keymeulen, L.E. Christensen, S.P. Schwenzer, R. Navarro-Gonzalez, J. Pla-GarcíÃ, S.C.R. Rafkin, Á. Vicente-Retortillo, H. Kahanpää, D. Viudez-Moreiras, M.D. Smith, A.-M. Harri, M. Genzer, D.M. Hassler, M. Lemmon, J. Crisp, S.P. Sander, R.W. Zurek, A.R. Vasavada, Background levels of methane in mars' atmosphere show strong seasonal variations. Science **360**, 1093–1096 (2018). https://doi.org/10.1126/science.aaq0131
15. S.E. Lauro, E. Pettinelli, G. Caprarelli, L. Guallini, A.P. Rossi, E. Mattei, B. Cosciotti, A. Cicchetti, F. Soldovieri, M. Cartacci, F. Di Paolo, R. Noschese, R. Orosei, Multiple subglacial water bodies below the south pole of mars unveiled by new MARSIS data. Nat. Astron. (in press)
16. R. Orosei, S.E. Lauro, E. Pettinelli, A. Cicchetti, M. Coradini, B. Cosciotti, F. Di Paolo, E. Flamini, E. Mattei, M. Pajola, F. Soldovieri, M. Cartacci, F. Cassenti, A. Frigeri, S. Giuppi, R. Martufi, A. Masdea, G. Mitri, C. Nenna, R. Noschese, M. Restano, R. Seu, Radar evidence of subglacial liquid water on mars. Science **361**, 490–493 (2018). https://doi.org/10.1126/science.aar7268
17. M.M. Sori, A.M. Bramson, Water on mars, with a grain of salt: local heat anomalies are required for basal melting of ice at the south pole today. Geophys. Res. Lett. **46**, 1222–1231 (2019). https://doi.org/10.1029/2018GL080985
18. J.L. Vago, F. Westall, A.J. Coates, R. Jaumann, O. Korablev, V. Ciarletti, I. Mitrofanov, J.-L. Josset, M.C. De Sanctis, J.-P. Bibring, F. Rull, F. Goesmann, H. Steininger, W. Goetz, W. Brinckerhoff, C. Szopa, F. Raulin, F. Westall, H.G.M. Edwards, L.G. Whyte, A.G. Fairén, J. Bridges, E. Hauber, G.G. Ori, S. Werner, D. Loizeau, R.O. Kuzmin, R.M.E. Williams, J. Flahaut, F. Forget, D. Rodionov, O. Korablev, H. Svedhem, E. Sefton-Nash, G. Kminek, L. Lorenzoni, L. Joudrier, V. Mikhailov, A. Zashchirinskiy, S. Alexashkin, F. Calantropio, A. Merlo, P. Poulakis, O. Witasse, O. Bayle, S. Bayón, U. Meierhenrich, J. Carter, J.M. GarcíÃ-Ruiz, P. Baglioni, A. Haldemann, A.J. Ball, A. Debus, R. Lindner, F. Haessig, D. Monteiro, R. Trautner, C. Voland, P. Rebeyre, D. Goulty, F. Didot, S. Durrant, E. Zekri, D. Koschny, A. Toni, G. Visentin, M. Zwick, M. van Winnendael, M. Azkarate, C. Carreau, ExoMars project team: habitability on early mars and the search for biosignatures with the ExoMars rover. Astrobiology **17**, 471–510 (2017). https://doi.org/10.1089/ast.2016.1533
19. M.S. Burgin, V. Stamenković, R.E. Grimm, D. Arumugam, R. Beauchamp, N. Barba, R. Manthena, D. Wright, B.H. Wilcox, K. Carpenter, C.D. Edwards, Searching for liquid water in the martian deep with TH2OR. American Geophysical Union, Fall Meeting 2019, abstract #P44B-02 (2019)
20. J.A. Grant, M.P. Golombek, S.A. Wilson, K.A. Farley, K.H. Williford, Al Chen, The science process for selecting the landing site for the 2020 mars rover. Planet. Space Sci. **164**, 106–126 (2018). https://doi.org/10.1016/j.pss.2018.07.001

21. G. Horneck, N. Walter, F. Westall, J.L. Grenfell, W.F. Martin, F. Gomez, S. Leuko, N. Lee, S. Onofri, K. Tsiganis, R. Saladino, E. Pilat-Lohinger, E. Palomba, J. Harrison, F. Rull, C. Muller, G. Strazzulla, J.R. Brucato, P. Rettberg, M.T. Capria, AstRoMap European astrobiology roadmap. Astrobiology **16**, 201–243 (2016). https://doi.org/10.1089/ast.2015.1441
22. B.L. Carrier, D.W. Beaty, M.A. Meyer, J.G. Blank, L. Chou, S. DasSarma, D.J. Des Marais, J.L. Eigenbrode, N. Grefenstette, N.L. Lanza, A.C. Schuerger, P. Schwendner, H.D. Smith, C.R. Stoker, J.D. Tarnas, K.D. Webster, C. Bakermans, B.K. Baxter, M.S. Bell, S.A. Benner, H.H. Bolivar Torres, P.J. Boston, R. Bruner, B.C. Clark, P. DasSarma, A.E. Engelhart, Z.E. Gallegos, Z.K. Garvin, P.J. Gasda, J.H. Green, R.L. Harris, M.E. Hoffman, T. Kieft, A.H.D. Koeppel, P.A. Lee, X. Li, K.L. Lynch, R. Mackelprang, P.R. Mahaffy, L.H. Matthies, M.A. Nellessen, H.E. Newsom, D.E. Northup, B.R.W. O'Connor, S.M. Perl, R.C. Quinn, L.A. Rowe, B. Sauterey, M.A. Schneegurt, D. Schulze-Makuch, L.A. Scuderi, M.N. Spilde, V. Stamenković, J.A. Torres Celis, D. Viola, B.D. Wade, C.J. Walker, R.C. Wiens, A.J. Williams, J.M. Williams, J. Xu, Mars extant life: what's next? Conference report. Astrobiology **20**, 785–814 (2020). https://doi.org/10.1089/ast.2020.2237

Chapter 13
Getting Ready for the SKA SETI

F. Schillirò and G. Umana

Abstract The Square Kilometre Array (SKA) in the next future will be able to reach the proper sensitivity to make it possible the interception of potential alien transmission at distance that is now certainly proved to host a big amount of exo-planetary systems; some of them are potentially able to host intelligent life. With applying Commensal Observation strategies and Cognitive and AI Computing, SKA, with the two stages of construction SKA1 and SKA2, could be considered a very good instrumentation to carry out the most important discovery of mankind.

13.1 A New Strategy for SETI Observation and Data Management

13.1.1 SKA Sensitivity for SETI

Historically, performing SETI experiments effectively with large telescopes has been technically and politically difficult.

A poor amount of time telescope allocated for targeted observation produced a large set of back-ends and dedicated electronic resources for piggy-back SETI observation, even if from a technical point of view an important question is still open: sensitivity. SKA will be the first world-class telescope ever constructed for which SETI observations are a conscious part of the design process and are available as a facility observing mode.

A the same time, SKA may be the first telescope capable of detecting truly Earth-like leakage radiation from nearby stars, allowing us our best chance at detecting another civilization with an artificial radio signature similar to our own.

In Fig. 13.1 is reported the Sensitivity of each component of the SKA to narrow-band transmitters at 15 pc, as compared with other facilities performing SETI searches over the same band, assuming a significance threshold $\sigma = 15$, bandwith

F. Schillirò (✉) · G. Umana
INAF-Osservatorio Astrofisico di Catania, via S. Sofia n. 78, 95123 Catania, Italy
e-mail: francesco.schilliro@inaf.it

S. Montebugnoli and A. Melis (eds.), *The Search for ExtraTerrestrial Intelligence*, Springer Proceedings in Physics 260, https://doi.org/10.1007/978-3-030-63806-1_13

$\Delta b = 0.5$ Hz and integration time $t = 10$ min. A transmitter is detectable if its EIRP is above the curve for a given telescope. So that a transmitter with an EIRP of 2×10^{20} erg/s (planetary radar) is detectable with all the telescope shown in figure, while a transmitter with an EIRP of 2×10^{17} erg/s (airport radar) is detectable only with SKA2 [1].

13.1.2 Commensal Observation Strategy and AI Computing

The serendipity nature of a SETI observation make it not always accepted within the scientific community, as targeted observation strategy, so that such kind of observations have less probability to be performed, without clear and unequivocal signs of intelligent and communicative life. This question will probably affect SKA telescope observation as well, so that a solution is the Commensal Observation Strategy and the Artificial Intelligence approach to Big Data (Fig. 13.2) [2].

In biology, commensalism is a symbiotic arrangement between mutualism and parasitism: it is the coexistence of two organisms in which one benefits and the other neither benefits nor suffers. It derives from the Latin 'cum mensa', "sharing a table," originally alluding to the sharing of food scraps in particular.

Commensalism is possible thanks to back-end technologies developed by a number of SETI programs performed all over the world: let us think to new beam-forming technologies and new back-ends with very high performance digitization and computing resources, that allows to use all the sensitivity theoretically provided by telescopes, in particular SKA.

In a five year commensal campaign, SKA1 could survey every star in its declination range within 60 pc more than ten thousand stars to a luminosity limit an order of magnitude fainter, ergs/s, over a larger band. Conservatively, in ten years SKA2 could survey every star within 60 pc to a luminosity limit equal to the EIRP of terrestrial aircraft radars over the entire terrestrial microwave window. Commensal observations will soon begin in the radio at the MeerKAT radio telescope in South Africa [3].

As large corporate investments are made in the commercial exploitation of Big Data, significant progress is being made in areas such as advanced data analytics, new visualisation methods and the introduction of advanced pattern recognition and feature extraction techniques.

These can greatly benefit radio astronomy in general and perhaps SETI searches in particular. As noted earlier, the likely rate of radio transients in the era of the full SKA will reach levels that are beyond the capacity of even large, well organized teams of radio astronomers.

Faced with an avalanche of data from all sides, the "brute force" approach of commercial data analytics, and especially the future promise of cognitive computing, could have a huge impact in terms of increasing the chances of serendipitous discovery—fields of research such as SETI, where human bias and other pre-conceptions may limit current efforts, stand to benefit enormously.

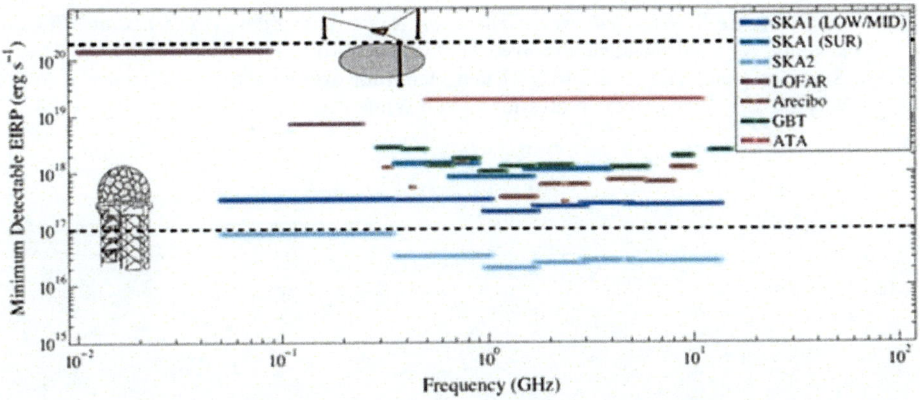

Fig. 13.1 Sensitivity of each component of the SKA to narrow-band transmitters at 15 pc, as compared with other facilities performing SETI searches over the same band. a transmitter with an EIRP of 2×10^{20} erg/s (planetary radar) is detectable with all the telescope shown in figure, while a transmitter with an EIRP of 2×10^{17} erg/s (airport radar) is detectable only wit SKA2

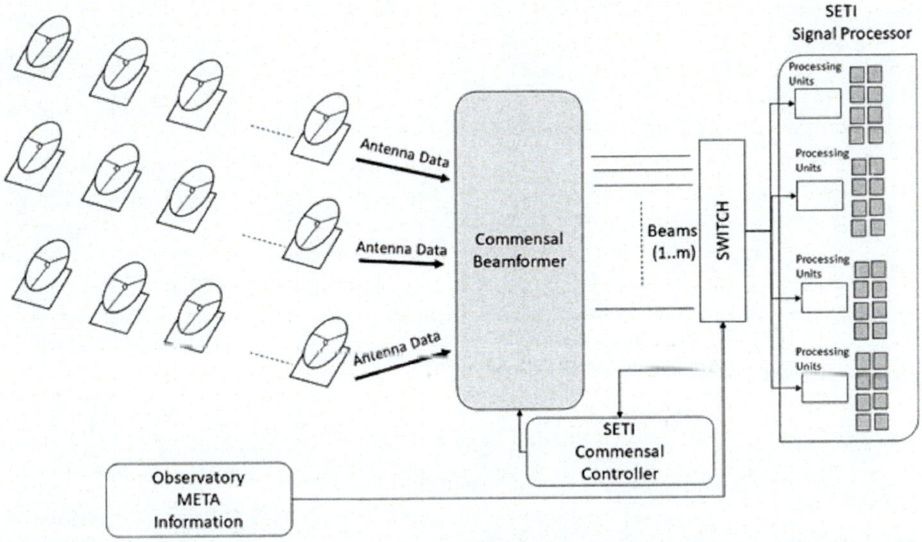

Fig. 13.2 Schematic diagram of Commensal Observation on the SKA: a SETI observer employs an independent beamforming capability to form phased array beams within the current primary field(s) of view based on observatory meta information, and directs the unprocessed digital voltage data for those beams to SETI signal processor

References

1. A. Siemion, *Searching for Extraterrestrial Intelligence with the Square Kilometre Array, Proceedings of Science*

2. M.A. Garret, *SETI Reloaded: Next Generation Radio Telescopes, Transients and Cognitive Computing*. Acta Astronautica, Received 31 December 2014
3. A. Siemion, et al., The breakthrough listen search for extraterrestrial intelligence, astro2020 activity, project or state of the profession (APC) White Paper

Chapter 14
SETI in Rocky Exoplanets: Narrowing the Search with Climate Models

Giovanni Vladilo, Laura Silva, Michele Maris, Giuseppe Murante, and Paolo Simonetti

Abstract The quest for inhabited worlds beyond the Solar System is focussed on rocky exoplanets, many of which are being discovered via transit and radial-velocity surveys. The detection of life in such planets via atmospheric biosignatures is challenging and ambient conditions that maximize the production and detectability of atmospheric biosignatures should be preferred in the selection of targets for spectroscopic observations. In this presentation we discuss how climate models that predict the temperature distribution on the planetary surface and the absorption properties of the planetary atmosphere can be used to narrow the search for exoplanets able to sustain multicellular organisms and hence, potentially, intelligent life.

14.1 Introduction

The study of the potential distribution of life in the Galaxy is one of the main goals of astrobiology and may provide effective strategies for the search of extraterrestrial intelligence (SETI). Rocky exoplanets are natural candidates in the quest for astronomical environments potentially able to host life. Based on the example of terrestrial life, the emergence of complex, multicellular life is a pre-condition for the development of neural connections, brains and, eventually, intelligent life. For this reason, in the first part of this presentation we discuss the physical conditions that are required to sustain multicellular life of terrestrial type on a planetary surface. We focus, in particular, on the ambient temperature, which plays a key role in the regulation of life processes [1]. Since the detection of life in exoplanets relies on the spectroscopic observation of atmospheric biosignatures [2], we also discuss the temperature limits that may maximize the generation of chemical constituents of

G. Vladilo (✉) · L. Silva · M. Maris · G. Murante
INAF - Osservatorio Astronomico di Trieste, Trieste, Italy
e-mail: giovanni.vladilo@inaf.it
URL: http://wwwuser.oats.inaf.it/vladilo/

P. Simonetti
Università degli Studi di Trieste, Trieste, Italy

S. Montebugnoli and A. Melis (eds.), *The Search for ExtraTerrestrial Intelligence*, Springer Proceedings in Physics 260, https://doi.org/10.1007/978-3-030-63806-1_14

biological origin. Searching for atmospheric biosignatures is one of the top priorities of exoplanetary observations that will be carried out with the next generation astronomical facilities in space (e.g. JWST) and on ground (e.g. E-ELT). The detection of atmospheric bio-signatures in rocky planets is extremely challenging and requires a careful pre-selection of suitable targets. In turn, this requires a modelization of the surface and atmospheric properties of the planet, starting from the modest amount of experimental data that can be obtained from the observational methods of exoplanets [3]. Modelling the surface conditions is essential to characterize the habitability of the planet and to understand if surface life, if present, can generate atmospheric signatures. Modelling the atmosphere is necessary not only to estimate the impact of atmospheric feedbacks on the climate and habitability, but also to calculate the optical depth of atmospheric biosignatures that could be detected with spectroscopic methods. The modelization of surface and atmospheric properties of exoplanets can be accomplished with the aid of dedicated climate models that we briefly discuss in the second part of this presentation.

14.2 Thermal Limits of Multicellular Life

Life has several requirements that may be used to define criteria of habitability. These requirements include, among others, the existence of suitable energy sources, physical conditions, protection from ionizing radiation, and an appropriate set of chemical constituents. The thermodynamical conditions that allow water to be present in liquid phase on the planetary surface are commonly used to define the liquid-water criterion which, with the aid of climate models, is applied to estimate the extension of the habitable zone (HZ) around planet-hosting stars [4, 5]. The liquid-water criterion provides temperature limits that can be parametrized as a function of surface atmospheric pressure to define a pressure-dependent HZ [6].

Beside its importance for the potential existence of liquid water, the ambient temperature can be used to set thermal limits of habitability based on the temperature dependence of biological processes [7]. Terrestrial life is characterized by thermal limits of survival, metabolism, and reproduction that are specific for different types of organisms [8]. Here we consider the thermal limits of multicellular organisms with active metabolism. We focus on multicellular organisms because they represent a necessary step along the evolutionary pathways that lead to the emergence of life with neural connections and brains, i.e. the type of life which is of interest for SETI. We focus on organisms with active metabolism because this is the only type of life that can generate a detectable chemical imprint in the exoplanetary atmosphere.

Among terrestrial organisms, poikilotherms are of special interest for setting thermal limits of habitability because their internal temperature depends directly on and varies with ambient temperature [1, 7]. Conversely, homeotherms do not provide straightforward limits of ambient temperature because they are able to stabilize their internal conditions over a broad range of external temperatures [9]. Homeotherms are interesting in the context of SETI, because a tight control of the internal body temperature seems to be essential for the functioning of brains. Since homeotherms emerged from Darwinian evolution of multicellular poikilotherms, we may say that the thermal limits of multicellular poikilotherms are relevant for all multicellular life, including homeotherms.

The approximate thermal limits of multicellular poikilotherms (plants, invertebrates and ectothermic vertebrates) with active metabolism are $0 \leq T(°C) \leq 50$ [8]. Quite interestingly, the same limits are also relevant for the biological production of atmospheric O_2 because the metabolism of the main O_2 producers (cyanobacteria and plants) drops outside this temperature interval [7]. It is hard to overemphasize the role of oxygen in this context, since the aerobic metabolism is much more efficient that anaereobic metabolism and the presence of significative amounts of atmospheric O_2 might be a necessary condition for the emergence of multicellular life in any planet [10].

The thermal limits $0 \leq T(°C) \leq 50$ are more stringent than the liquid-water temperature range commonly adopted in studies of habitability. These stringent limits can be used to narrow the search of optimal targets for the SETI program.

14.2.1 *How Universal Are the Thermal Limits of Terrestrial Life?*

A comprehensive mechanistic understanding of the effects of temperature on biological processes is still lacking. In spite of this, the physical nature of the main mechanisms of thermal response at work in terrestrial life suggest that the same mechanisms would also be at work in other forms of chemical life.

At the molecular level, there are strong indications that life processes based on genetic and catalytic molecules require the existence of a network of hydrogen-bond interactions [11]. Among cosmically abundant elements and molecules, CNO elements and water have unique capabilities to form molecular groups that can interact via hydrogen bonds. Therefore, the basic aspects of terrestrial biochemistry (CHON elements and water) are likely to be universal for any type of life life based on genetic and catalytic molecules.

For water-based life, the water freezing point $T = 0\,°C$, which is almost independent of the ambient pressure, is likely to be a universal lower bound, because frozen water would hamper the mobility of biomolecules and, in addition, would make impossible for molecular motors [12] to harvest the kinetic energy of Brownian motion [7]. A universal upper bound is set by the temperature at which thermal energy denatures the molecular structures most sensitive to heat. Since intramolecular hydrogen bonds are essential for shaping molecular structures, the low binding energy of hydrogen bonds can be used, in principle, to set universal upper limits to the temperature of life processes [11].

At the supramolecular level, the thermal tolerance must narrow as complexity increases, because the number of molecular structures potentially limiting the thermal tolerance increases with increasing complexity of the organism. The study of metabolic processes casts light on the supramolecular mechanisms of thermal response, suggesting that aerobic metabolism sets tighter thermal limits than other high-level functions typical of multicellular organisms [13].

The above considerations suggest that any form of multicellular, aerobic life will be characterized by stringent thermal limits due to the combined constraints that originate at the molecular level and at higher levels of structural and functional complexity. A better understanding of these mechanisms may eventually lead to the definition of universal thermal limits for specific forms of life. For the moment, in lack of better indications, we adopt the thermal limits $0 \leq T\,(°\mathrm{C}) \leq 50$, representative of terrestrial multicellular poikilotherms, as a criterion of long-term habitability of complex life outside Earth. The fact that such limits are shared by multicellular poikilotherms emerged from independent evolutionary pathways on Earth is consistent with this assumption.

14.3 Modelling the Surface Temperature of Rocky Exoplanets

Based on the above discussion we use the ambient temperature as a tool for assessing the capability of a planet to host complex life. The surface temperature of exoplanets can be calculated by inserting observational data in dedicated climate models. Unfortunately, only a small amount of data can be measured for individual exoplanets [3]. The data obtained from the transit and radial velocity surveys may include planetary structural parameters (radius, mass), orbital parameters (semi-major axis, eccentricity), and properties of the host star (luminosity, spectral type, chemical composition, and age). Other planetary quantities that impact the climate, but are currently not measurable (e.g. rotation period, axis tilt, geography, surface pressure, atmospheric composition) must be treated as free model parameters.

Climate models for exoplanets must deal with the intrinsic complexity of the climate system, characterized by different components, processes, feedbacks and time scales [14]. In addition, exoplanet climate models must be adapted to simulate conditions which are not treated in Earth climate models. Given these difficulties, a hierarchy of climate models should be employed in exoplanet studies. Models of higher complexity, tested with well-known climates of rocky planets (present Earth, Mars, Venus, paleo-Earth), should be used to validate models of lower complexity. Once properly validated, models of low complexity can be used to explore the huge parameter space that characterizes the stellar, orbital and planetary properties of rocky exoplanets.

Classic studies of the HZ have used single atmospheric column calculations, with a simplified, radiative-convective treatment of the vertical transport and an albedo representative of the mean planetary albedo [4, 5]. With an idealized treatment of the latitudinal energy transport, Energy Balance Models (EBMs) can simulate latitudinal and seasonal variations of the surface temperature [15] and have been applied to studies of planetary habitability [16]. By incorporating single column, radiative-convective calculations and a schematic treatment of the clouds in classic EBMs, one obtains 2D (vertical and latitudinal) models with seasonal dependence of the surface temperature [6, 17]. A refinement of this last type of model is the Earth-like planet surface temperature model (ESTM), which incorporates a physically-based description of the meridional transport validated with models of higher complexity [18]. The ESTM provides fast estimates of the zonal surface temperature, $T_s = T_s(\varphi, t)$, as a function of latitude, φ, and time, t. The temperature distribution predicted in this way can be used to characterize the habitability of exoplanets.

14.3.1 A Temperature-Dependent Index of Complex-Life Habitability

To introduce a quantitative index of habitability we define a temperature interval, (T_1, T_2), which we consider optimal for the maintainance of multicellular life and the production of atmospheric biosignatures. From a modelization of the zonal surface temperature, $T_s = T_s(\varphi, t)$, such as that provided by the ESTM, we then calculate a habitability function

$$H(\varphi, t) = \begin{cases} 1 & \text{if } T_1 \leq T_s(\varphi, t) \leq T_2 \\ 0 & \text{otherwise} \end{cases} . \tag{14.1}$$

By averaging $H(\varphi, t)$ in φ (weighting latitude zones according to their area) and t (over one orbital period), we obtain the index of mean planetary habitability

$$h = \frac{\int_{-\frac{\pi}{2}}^{+\frac{\pi}{2}} d\varphi \int_0^P dt \, [H(\varphi, t) \cos\varphi]}{2P} . \tag{14.2}$$

The choice of the temperature limits (T_1, T_2) is a critical point of this operational definition of habitability. Based on the thermal limits for complex life discussed above, we adopt $T_1 = 0\,°\text{C}$ and $T_2 = 50\,°\text{C}$ in Eq. (14.1) and we call h_{050} the index calculated from Eq. (14.2) with these limits. The index h_{050}, normalized between 0 and 1 by construction, provides a quantitative estimate of the mean planetary habitability for multicellular life. Moreover, the temperature range that maximizes h_{050} is the same range in which terrestrial cyanobacteria and plants produce oxygen [7]. This is important because oxygen is a potential atmospheric biosignature and is probably necessary for the emergence of complex life [10].

In addition to the above considerations, driven by biological arguments, the index h_{050} presents an important advantage in terms of climate calculations: the low value of the upper thermal limit, $T_2 = 50\,°\text{C}$, avoids the necessity to perform climate calculations in a regime of high temperatures that may lead to the onset of runaway greenhouse instability [4]. This regime, characterized by high water vapour content, is extremely hard to model even with state-of-the-art, 3D climate models [19].

14.3.2 Applications to Studies of Habitability for Complex Life

By using the methodology described above we can estimate the index h_{050} for a broad range of factors that affect the planetary climate. Studies performed with the ESTM show that the parameters that most heavily impact $T_s(\varphi, t)$ and h_{050} are the insolation, the atmospheric properties (pressure and composition) and the albedo (surface and clouds) [18]. The index h_{050} can be applied to individual exoplanets or to perform statistical studies of exoplanetary habitability.

Kepler-452b is an example of an Earth-size planet ($R = 1.63\,R_\oplus$) in the HZ of a sun-like star [20] that has been studied with the ESTM [21]. Under the assumption of a rocky-dominated nature, $T_s(\varphi, t)$ and h_{050} were computed for a broad range of climate factors. Even if the insolation of Kepler-452b is only 10% larger than that received by Earth, the constraints of habitability for complex life are very stringent. For most choices of parameters, habitable solutions with $h_{050} > 0.2$ are only found if the CO_2 partial pressure is $pCO_2 < 0.04\,\mathrm{bar}$. At this limiting value of pCO_2, the planet is habitable only if the total pressure is $p < 2\,\mathrm{bar}$. In all cases, the habitability index h_{050} drops to zero if the orbital eccentricity is $e > 0.3$. Changes of rotation period and axis tilt affect h_{050} due to their impact on the equator–pole temperature difference, which affects the possible existence of polar caps. Variations of h_{050} resulting from the luminosity evolution of the host star were estimated with the aid of stellar evolutionary tracks [22]. Only a small combination of parameters yields habitability-weighted lifetimes > 2 Gyr, sufficiently long to develop atmospheric biosignatures still detectable at the present time [21]. This study illustrates the importance of exploring the parameter space of climate factors in order to assess the potential of individual planets to host complex life.

Thanks to the flexibility of the ESTM, it is possible to run a large number of climate simulations and perform statistical studies of habitability. An example of this statistical approach is the study of the bistability of the planetary climate as a function of the initial conditions of the simulations. An intriguing result of this type of study is that the planetary conditions that support climate bistability are remarkably similar to those required for the sustenance of multicellular life on the planetary surface [23]. The statistical approach can also be used to build up multi-parameter HZs by calculating h_{050} as a function of the planet insolation, S, and other climate-impacting parameters [7, 24, 25]. At variance with the classic HZ, thanks to the low value of the upper thermal limit ($T_2 = 50\,^\circ\mathrm{C}$), the inner edge of the complex-life HZ can be calculated without simulating the conditions that drive the runaway greenhouse instability.

A parameter of special interest for building up the HZ for complex life is the atmospheric columnar mass, $N_{atm} = p/g$, where p and g are the surface pressure and gravitational acceleration, respectively. The atmospheric columnar mass has a strong impact on the climate because it affects the energy transport both along the surface (horizontal transport) and between the surface and the top of the atmosphere (vertical transport). The atmospheric columnar mass also acts as a protective shield for life potentially present on the planetary surface, by absorbing and degrading cosmic rays of stellar or Galactic origin. For a planet with Earth-like characteristics (including the magnetic field), the surface dose of secondary particles of cosmic rays exceeds 100 mSv/year when $N_{atm} < 300$ g/cm^2 [26]. By displaying h_{050} as a function of S and N_{atm}, it is possible to build up an atmospheric mass habitable zone (AMHZ) for complex life [7]. The calculations of the AMHZ can be repeated for different atmospheric compositions, showing the impact of greenhouse gases, such as CO_2, on the location of the HZ. The AMHZ calculated with the index h_{050} is generally narrower than the classic HZ, providing tight constraints for the search of exoplanets capable of hosting multicellular life. By decreasing N_{atm}, the inner edge of the HZ gets closer to the star due to the decrease of the greenhouse heating of the surface. However, this effect becomes negligible at low values of N_{atm} and below 300 g/cm^2 the only net result is a significant rise of the surface dose of radiation [7, 26].

14.4 Conclusions

By investigating the thermal limits of multicellular life it is possible to introduce an operational definition of complex-life habitability that can be applied to SETI-oriented studies of exoplanets. Based on experimental data of terrestrial life, the temperature interval $0 \leq T(°\mathrm{C}) \leq 50$ is suitable for the emergence of complex life and for the biological generation of atmospheric O_2, a biosignature potentially detectable with spectroscopic observations of exoplanetary atmospheres. With the aid of dedicated climate models, it is possible to predict the surface temperature distribution of rocky exoplanets by combining observational data with a parameterization of the climate factors that are currently unconstrained by observations. By modelling the surface temperature we can assess which range of parameter space is suitable for the sustenance of complex life and the detection of atmospheric biomarkers.

For future applications of this methodology it is desirable to upgrade climate models to be able to simulate a broad range of stellar, orbital and planetary conditions, including the climate impact of biological feedbacks. By adopting a multi-parameter approach to the study of planetary habitability it will be possible to broaden the concept of habitable zone [24, 25, 27]. To assess the universal validity of the temperature limits deduced from the properties of terrestrial life, we need to improve our understanding of the physical mechanisms that govern the thermal response of life processes. Statistical studies of exoplanets with properties suitable to sustain complex life can be applied to study the potential distribution of intelligent life in the Galactic Habitable Zone [28].

Acknowledgements We thank the Italian Space Agency for co-funding the Life in Space project (ASI N. 2019-3-U.0).

References

1. H. Precht, J. Christophersen, H. Hensel, W. Larcher, *Temperature and Life* (Springer, Berlin, Heidelberg, 1973)
2. E.W. Schwieterman, N.Y. Kiang, M.N. Parenteau et al., Exoplanet biosignatures: a review of remotely detectable signs of life. Astrobiology **18**, 663 (2018)
3. M. Perryman, *The Exoplanet Handbook*, 2nd edn. (Cambridge University Press, 2018). ISBN: 9781108419772
4. J.F. Kasting, D.P. Whitmire, R.T. Reynolds, Habitable zones around main sequence stars. Icarus **101**, 108 (1993)
5. R.K. Kopparapu, R. Ramirez, J.F. Kasting et al., Habitable zones around main-sequence stars: new estimates. Astrophys. J. **765**, 131 (2013)
6. G. Vladilo, G. Murante, L. Silva et al., The habitable zone of earth-like planets with different levels of atmospheric pressure. Astrophys. J. **767**, 65 (2015)
7. L. Silva, G. Vladilo, P.M. Schulte et al., From climate models to planetary habitability: temperature constraints for complex life. Int. J. Astrobiol. **16**, 244 (2017)
8. A. Clarke, The thermal limits to life on earth. Int. J. Astrobiol. **13**, 141 (2014)
9. J. Ruben, The evolution of endothermy in mammals and birds: from physiology to fossils. Ann. Rev. Physiol. **57**, 69 (1995)
10. D.C. Catling, C.R. Glein, K.J. Zahnle, C.P. McKay, Why O_2 is required by complex life on habitable planets and the concept of planetary "oxygenation time". Astrobiology **5**, 415 (2005)
11. G. Vladilo, A. Hassanali, Hydrogen bonds and life in the universe. Life **8**, 1 (2018). https://doi.org/10.3390/life8010001
12. C. Bustamante, D. Keller, G. Oster, The physics of molecular motors. Acc. Chem. Res. **34**, 412 (2001)
13. P.M. Schulte, The effects of temperature on aerobic metabolism: towards a mechanistic understanding of the responses of ectotherms to a changing environment. J. Exp. Biol. **218**, 1856 (2015)
14. A. Provenzale, Anthropocene—natural and man-made alterations of the earth's fragile equilibrium. Rend. Fis. Acc. Lincei (2013). https://doi.org/10.1007/s12210-013-0268-7
15. G.R. North, R.F. Cahalan, J.A. Coakley, Energy balance climate models. Rev. Geophys. Space Phys. **19**, 91 (1981)
16. D.S. Spiegel, K. Menou, C.A. Scharf, Habitable climates. Astrophys. J. **681**, 1609 (2008)
17. D.M. Williams, J.F. Kasting, Habitable planets with high obliquities. Icarus **129**, 254 (1997)
18. G. Vladilo, L. Silva, G. Murante, L. Filippi, A. Provenzale, Modeling the surface temperature of earth-like planets. Astrophys. J. **804**, 50 (2015)
19. J. Leconte, F. Forget, B. Charnay et al., Increased insolation threshold for runaway greenhouse processes on earth-like planets. Nature **504**, 268 (2013)
20. J.M. Jenkins, J.D. Twicken, N.M. Batalha, et al., Discovery and validation of Kepler-452b: a 1.6 $R_\oplus$ super earth exoplanet in the habitable zone of a G2 star. Astron. J. **150**, 56 (2015)
21. L. Silva, G. Vladilo, G. Murante et al., Quantitative estimates of the surface habitability of Kepler-452b. Monthly Notices R. Astron. Soc. **470**, 2270 (2017)
22. A. Bressan, P. Marigo, L. Girardi et al., PARSEC: stellar tracks and isochrones with the PAdova and TRieste stellar evolution code. Monthly Notices R. Astron. Soc. **427**, 127 (2012)
23. G. Murante, A. Provenzale, G. Vladilo et al., Climate bistability of earth-like exoplanets. Monthly Notices R. Astron. Soc. **492**, 2638 (2020)
24. G. Murante, M. Maris, E. Palazzi et al., Estimating planetary habitability in a wide parameter-space modelling study. EGU General Assembly Conf. Abstracts **6090** (2018)

25. G. Vladilo, Exploring the habitability of exoplanets with climate models. Memorie della Societa Astronomica Italiana **90**, 648 (2019)
26. D. Atri, B. Hariharan, J.-M. Griemeier, Astrobiology **13**, 910 (2013)
27. R. Ramirez, D.S. Abbot, Y. Fujii et al., Habitable zone predictions and how to test them. Bullet. Am. Astron. Soc. **51**, 31 (2019)
28. I.S. Morrison, M.G. Gowanlock, Extending galactic habitable zone modeling to include the emergence of intelligent life. Astrobiology **15**, 683 (2015)